# MULTISENSOR FUSION

## A Minimal Representation Framework

# SERIES IN INTELLIGENT CONTROL AND INTELLIGENT AUTOMATION

**Editor-in-Charge:**   Fei-Yue Wang
*(University of Arizona)*

---

Series in
Intelligent Control and Intelligent Automation
Vol. 11

# MULTISENSOR FUSION
## A Minimal Representation Framework

**Rajive Joshi**
*Real-Time Innovations Inc., USA*

**Arthur C Sanderson**
*Rensselaer Polytechnic Institute, USA*

**World Scientific**
*Singapore • New Jersey • London • Hong Kong*

*Published by*

World Scientific Publishing Co. Pte. Ltd.

P O Box 128, Farrer Road, Singapore 912805

*USA office:* Suite 1B, 1060 Main Street, River Edge, NJ 07661

*UK office:* 57 Shelton Street, Covent Garden, London WC2H 9HE

**British Library Cataloguing-in-Publication Data**
A catalogue record for this book is available from the British Library.

MULTISENSOR FUSION: A MINIMAL REPRESENTATION FRAMEWORK

ISBN 981-02-3880-0

Printed in Singapore by Uto-Print

To my father Sri. Liladhar Joshi
and
To the memory of my grandparents
Sri. Laxmidutt Joshi and Smt. Ganga Joshi
- Rajive

# Preface

Multisensor fusion has emerged as a central problem in the development of robotic systems where interaction with the environment is critical to the achievement of a given task. As applications become more complex, the selection of model, parameters, and data subsamples is more difficult to determine a priori, and therefore the availability of a consistent framework across a wide variety of problem domains is important.

The primary focus of this book is to describe a methodology for addressing model selection and multisensor data fusion issues, originally developed as part of the first author's doctoral thesis work under the guidance of the second author. This book is divided into three parts. In Part 1, we present a broad survey of the literature on multisensor integration, multisensor fusion techniques, and applications in object recognition using multiple sensors. In Part 2 of the book, we describe the minimal representation size framework for multisensor fusion and model selection, and develop generic environment and sensor models, minimal representation encodings and their properties, and templates for generic multisensor fusion algorithms. In Part 3, we apply the abstract framework to multisensor object recognition problems, and describe laboratory experiments and practical experiences using the minimal representation size fusion methodology.

The general framework for multisensor fusion and model selection uses representation size (description length) as a *universal sensor independent yardstick* to choose (a) the model class and number of parameters from a library of environment models, (b) the model parameter resolution and data scaling based on the sensor resolution and accuracy, (c) the subset of observed data features which are modeled (and are therefore used in the environment model parameter estimation) based on sensor precision, and **(d)** the correspondence which maps data features to model features. The sensor precision, controls the balance between the number of modeled and unmodeled data features, and determines the data subsampling. Error residuals of subsampled data features from different sensors, "scaled" on the basis of sensor accuracy, and "weighted" on the basis of sensor resolution, are used to define the environment model parameters according to the sensor constraint equations.

The minimal representation size criterion may be used for the environment model parameter estimation itself, or alternatively a more traditional statistical estimation method may be used. The search for the best interpretation may be conducted using **(1)** polynomial time algorithms using constraining data feature sets to instantiate environment models, or **(2)** evolution programs which use principles of natural selection to reproduce a population of interpretations.

The application of the framework to object recognition focuses on solving the object identification, localization, and matching problems in two and three dimensional workspaces, using vision, touch, and grasp sensors. In the laboratory experiments, tactile data obtained from the finger-tips of a robot hand, while it is holding an object in front of a camera, is fused with the vision data from the camera, to determine the object identity, pose, and the touch and vision data correspondences. The touch data is incomplete due to required hand configurations, while nearly half of the vision data is spurious due to the presence of the hand in the image. Using either sensor alone results in ambiguous or incorrect interpretations, and multisensor fusion is necessary to consistently find the correct interpretation.

The experiments use a *differential evolution* program to search for the best interpretation. As developed in this work, the search problem is difficult because we have posed it as a broad search over many general pose configurations (local minima), and have chosen not to impose heuristic constraints to simplify the search. In practice, there are many such heuristics which may be imposed for specific problems, and they would often improve the execution time of the search. However, in this study the focus has been on the nature of the representation itself and the associated search algorithms, rather than on building a practical system. The understanding and performance of these measures and algorithms provides the basis for further improvement in practical systems.

The laboratory experiments demonstrate the automatic selection of environment model class (object identity), the environment parameter values (object pose), and the sensor data correspondences (touch and vision correspondences), in the minimal representation size framework for multisensor fusion and model selection.

In practice, robotic manipulation using multisensor data must be done in real-time, and it became clear that the general evolutionary algorithms are too slow to estimate pose at sampling speeds for continuous motion. In a real-time system, one would use the generalized algorithm described here to initialize and occasionally update the pose estimate, while a much faster real-time update algorithm based, for example, on a Kalman filter would provide real-time feedback.

The methodology presented in this book is a general approach to fusion which is not restricted to geometric pose estimation, and in fact may be applied to a variety of problems in model identification from a wide perspective, including model-based identification of parameters from noisy data and prioritization in noisy data sets. The extension of this approach to many different types of models and estimation problems is an important element of this methodology and its use for general parametric model estimation with noisy data is open for further research.

### Acknowledgements

We would like to thank World Scientific publishers, and in particular our editor Jeremiah Kwok and our series editor Professor Feiyue Wang, for selecting this work to be published as part of the series on intelligent control and intelligent automation.

The first author would like to express his gratitude to his coauthor and thesis advisor, Professor Art Sanderson, who guided him though his doctoral work, which forms the basis for this book. We also wish to acknowledge the members of the doctoral committee, including Professor George Nagy, Professor Harry Stephanou, and Professor Ellen Walker, for scrutinizing this work, and for their valuable comments, suggestions, and encouragement. We would also like to acknowledge support for this work, from the New York State Center for Advanced Technology (CAT) in Automation, Robotics, and Manufacturing at Rensselaer Polytechnic Institute.

The first author would also like to thank his parents, his brother Himanshu, and his wife Pammy for their love, encouragement, and support through the different stages of this work.

# Contents

# Chapter 1

# Introduction

## 1.1 Motivation

Sensory data processing and interpretation is a very basic characteristic of intelligent beings—one that empowers them to interact with their environment. Intelligent beings seem to have an innate ability to "fuse" multiple sources of information and integrate them into their decision making process. In our day-to-day experience we fuse visual, auditory, olfactory, tactile, and taste senses without any conscious effort. It appears that if we are to build an "intelligent machine" capable of interacting with its environment and making autonomous or semi-autonomous decisions, multisensor data fusion and interpretation technology would be crucial to its success. However, it turns out that multisensor data fusion and interpretation is a particularly difficult task, one that has captivated researchers since the turn of the century, and one on which the last word is yet to be written.

Not surprisingly therefore, multisensor fusion has emerged as a central problem in the development of robotic systems where interaction with the environment is critical to the achievement of a given task. Manipulation of objects in uncertain positions and geometries, and navigation through uncertain environments are typical examples.

Applications of multisensor data fusion and interpretation technology include industrial tasks such as materials handling, parts fabrication, inspection, assembly, and defense systems, such as multisensor multitarget tracking.

In this book, we focus on multisensor fusion and interpretation problems commonly encountered in robotics and industrial automation settings.

Fig. 1.1 The *Anthrobot* five-fingered hand holding an object in the field-of-view of a fixed camera. The contact data obtained from tactile sensors mounted on the fingertips is fused with the processed image data obtained from the camera, to estimate the position and orientation of the object.

These problems are characterized by multiple sensors observing a common environment, which may be partially or completely unknown. The observations from the various multiple sensors are fused to build a parameterized environmental model, which may then be used in manipulation and planning decisions. The environmental model structure itself is chosen from a library of parametric environmental model structures.

Although our motivation stems from a desire to build more capable autonomous and semi-autonomous machines, the multisensor fusion methodology developed in this book is applicable more broadly, including system identification and process control applications.

## 1.2   The Problem

Figure 1.1 shows an example of multisensor fusion used to guide manipulation of an object, taken from our laboratory setup. The robot may have picked the object from a parts-bin, for building a complex assembly. The *Anthrobot* [4] five-fingered hand grasps an object, and senses the contact points with the surface of the object using tactile sensors. The tactile sensors extract touch position and approximate surface normal in the kinematic reference frame of the hand. In addition, a CCD camera views the position of the same object and extracts vertex/edge features of the object image. Both the tactile features and the visual features are related to the position and orientation ("pose") of the object, and in practice we wish to combine these two sources of information to improve our ability to accurately manipulate the object. The fusion of the tactile and image feature data is used to derive an improved estimate of the object pose which guides the manipulation. In general, the object shape must also be identified, from a library of possible object shapes.

This example is representative of a broader class of problems, where data from multiple sensors is fused to deduce an interpretation in terms of an environment model. This class of problems is characterized by the following:

- **Library of Environment Models:** We can choose from a number of model structures and model parameters. In the above example, a *rigid* object model is characterized by a geometrical shape description, and six pose parameters. An *articulated* object model would include additional parameters for the extra degrees of freedom.

- **Multiple Sensor Models:** Different sensors may provide qualitatively diverse types of data about a common environment. Each observed datum may be viewed as providing a constraint on the possible interpretations of the environment, and diverse constraints must be combined in some "meaningful" fashion. In the above example, the tactile sensor data may be viewed as imposing planar constraints on the object faces. The vision sensor data may be regarded as imposing perspective projection constraints on the object vertices.

Each sensor can be further characterized as follows:

  - **Uncertainty:** We can choose among various uncertainty models to describe the observation errors. Furthermore, outliers in the observed data must be adequately characterized. In the above example, the tactile data errors may be described using a probability distribution. The vision data features are extracted by highly non-linear filtering and sampling of the raw data, and therefore may not be adequately characterized using a probability distribution. Instead, alternative uncertainty models, based on the camera accuracy and precision may be more appropriate (Section 6.3.1 on page 81).

  - **Correspondence:** Observed data features are related to environment model features according to some constraint relations. Their correspondence is often unknown, and must be determined. In the example above, for tactile data, several contact points may correspond to an object face, and therefore the correspondence between contact points and object faces is "many-to-one". For vision data at most one image vertex may correspond to an object vertex, and therefore the correspondence between the image vertices and the object vertices is "one-to-one".

– **Registration/Calibration:** In practice, observed data, are expressed in *local* sensor measurement coordinates, and must be related to a common or *global* frame of reference for fusion to take place. Often, an environment model in a global reference frame can be mapped into the local sensor measurement frames. This reference frame alignment, or registration, often depends critically on the sensor *calibration*. In the above example, the tactile data registration depends critically on good kinematic calibration of the robot hand-arm system. The vision data registration depends critically on good calibration of the intrinsic and extrinsic camera parameters.

In a typical multisensor fusion problem in robotics, such as the tactile–visual example shown in Figure 1.1, we can choose from a number of environment model structures, environment model parameters, and uncertainty models. The uncertainty and the registration/calibration models are often chosen *a priori*, whereas the sensor constraints can be obtained from physical modeling. However, the integration of a variety of sensors in a robotic system, and the use of that system in a complex environment makes the a priori choice of the environment model structure, parameters, and correspondence difficult. The object in Figure 1.1 may belong to some library of parameterized object models with corresponding choices of data scaling and data subsample selection as precursors to the pose estimation itself. In this book, we will address three important model selection issues for multisensor fusion:

(1) **Environment model class selection:** What is the environment model class? How many parameters are required to specify it? What is the parameter resolution? Thus, which object is it, and what form should be used for the object and pose description? The fusion scheme should be able to discriminate between different model classes and select the "best" model structure and number of parameters.

(2) **Environment model parameterization and Data scaling:**
What are the values of the environment parameters? How should
the data from different sensors be scaled to determine these parameter values? Thus, what environment model parameters and
sensor constraints should be used to define the pose? How should
the scaling of data from different sensors be defined? The sensor
data should be "weighted" consistently in the fusion scheme; sensors with greater uncertainty should have a smaller influence in
choosing the environment model parameters.

(3) **Data subsample selection:** Which data features are used to determine the environment model parameters? What subset of the
data is consistent in the definition of the pose for a given estimator? What data features should be considered outliers? The fusion
scheme should be able to reject outliers, and adequately label the
correspondences.

## 1.3 The Approach

Much of the recent progress in multisensor fusion [2, 73, 76, 112, 119, 123,
125] for robotics has been based on the application of existing statistical
tools to **(a)** estimate the position of objects with known geometric models [50, 143, 182], **(b)** estimate the parameterized shape of an object from
sensor information [5, 19, 52, 155, 178], and **(c)** estimate a probability distribution of objects or object features (e.g. surfaces) based on sensor models [53]. Particularly in the field of mobile robotics where an environment
can often be defined by two-dimensional maps of sparsely spaced obstacles,
estimation methods such as maximum likelihood and Kalman filters are
successfully applied to estimate the pose of objects and obstacles and have
become integral to many prototype mobile robot platforms [130]. As new
sensing methods [39] are introduced, these methods can often be directly
applied, or applied after attention to the creation of adequate probabilistic
models and approximations.

Whenever statistical models such as these are used, the selection of the
model and the number of model parameters is a critical step in the process.
Historically, this problem of model selection has been explored in a number
of classic statistical modeling domains, for example, autoregressive models
and cluster models, and these efforts have lead to a number of proposed

approaches to statistical model selection [3, 47, 175, 206]. These methods have in common the recognition of tradeoffs between model characteristics such as size (e.g. number of model parameters), parameter resolution, or optimal data subsampling and the degree of model fit (e.g. mean square error).

The *minimal representation size* (MRS) criterion [12, 30, 113, 142, 165, 177, 183] used in this work, more commonly known as *minimum description length* in the literature, belongs to this class of model selection methods. The MRS approach has been applied to a number of related problems in attributed image matching [173], shape matching [97], density estimation [13], and model-based multisensor fusion [95, 98–104]. The model selection problem is complementary to the estimation process itself and is intended to choose an effective combination of model structure and estimation method for a given class of problems.

The minimal representation size criterion chooses the minimal overall data representation, and leads to a choice among alternative models which trades-off between the size of the model (e.g. number of parameters) and the representation size of the (encoded) residuals, or errors. Intuitively, the "smaller", less complex, representation is chosen as the preferred model for a given estimator. In practice, a major advantage of this approach is the attainment of consistent results without the introduction of problem specific heuristics or arbitrary weight factors. In addition, the use of an information based criterion provides a type of "universal yardstick" for sensor data from many disparate sources, and therefore supports efficient implementation of these methods to new domains.

We develop a formulation of the minimal representation size criterion for a class of multisensor fusion problems. Expressions for representation size are derived based on the definition of constraints relating observed data features to the environment model. Within these constraint relations, the minimal representation size is used to choose the scaling for data representation, and the subsampling of the data set which results in the most consistent interpretation of the multisensor data.

The problem of tactile–visual fusion, described in Section 1.2 on page 2, is used to illustrate this methodology, where multiple tactile sensing points from separate finger contacts are combined with visual features from a single camera to determine the pose of an object model.

### 1.3.1 *Detailed Overview*

The minimal representation approach is based on the principle of building the shortest length program which reconstructs the observed data. The *representation size* or description length of an entity (Chapter 5) is defined as the length of the shortest length program that reconstructs the entity.

In the minimal representation size framework for multisensor fusion, shown in Figure 1.2, a model-based encoding scheme is used, in which the data is thought to be arising from one of the several available models in a model library. The environment model library may be regarded as the system's "knowledge" of the environment, containing models which may differ in structure and number of parameters. The representation size of an instantiated environment model is the smallest number of bits needed to describe the parameters and structure relative to the environment model library.

The observed data is encoded by specifying an instantiated environment model and the deviations or *residuals* of the observed data from that instantiated model. Each sensor is modeled by a general constraint equation and accompanying boundary conditions, which define a constraint manifold in the sensor measurement space. Observed sensor data populates the measurement space according to these constraints. Given a model of sensing errors, i.e., an uncertainty model, observed data can be encoded with respect to these constraint manifolds. The correspondence of the observed datum to the constraint manifolds is often unknown, and is encoded as an additional term in the sensor representation size.

The data representation size can be formally defined for a diverse variety of sensors, and may be regarded as the "information", in bits, contained in the observed data. Sensors may differ greatly in type of uncertainty models available to describe the observation errors. Often observation errors may be characterized or approximated by probability distributions. For such sensors, observed data is encoded using optimal coding schemes that achieve the *best data compression*. For some types of measurements such approximate distributions may become unreliable. In those cases, it is possible—and more natural—to use the sensor *accuracy* and *precision* directly as a model of the sensor uncertainty. For such sensors, observed data is encoded using *integer coding schemes* that give the shortest codewords.

The overall total representation size is obtained by adding the model representation size and the data representation size for each sensor. It in-

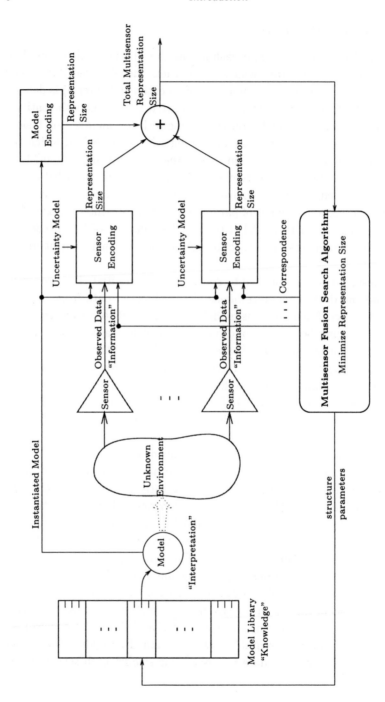

Fig. 1.2   Overview of the minimal representation size framework for multisensor fusion and model selection.

cludes terms for a stored environment model, a reconstruction algorithm, and encoded residuals. The "best interpretation" is an instantiated environment model, specified by its structure and parameters, that minimizes the overall total representation size. A *multisensor fusion algorithm* searches for the minimal representation size environment model and data correspondences, in the space of all instantiated models and possible correspondences (Figure 1.2).

The total representation size may be viewed as the smallest number of bits needed to transmit observed data from one point to another, assuming both the transmitter and the receiver have the same knowledge (environment model library), and a computer (Universal Turing Machine) to process the bits. Thus, representation size has the physical meaning as the most compact description of observed data. It gives us a *sensor independent universal yardstick*, which can be formally defined for a diverse variety of sensors, and provides a *meaningful* way of incorporating unknown correspondences of differing types and environment models with differing structure and parameters, in a multisensor fusion framework.

### 1.3.2 *Polynomial Model Order Selection Example*

For a better intuitive understanding of the minimal representation size approach, consider a polynomial curve fitting example as shown in Figure 1.3. The $N = 11$ data points $\mathbf{Z} = \{\mathbf{z}_i\}_{i=1}^N$ are generated from a second order polynomial by adding Gaussian noise:

$$\mathbf{z}_i = 3x_i^2 + 2x_i + 45 + \mathcal{N}_i(0, 15^2)$$

where $\mathcal{N}_i(0, \sigma^2)$ is a Gaussian noise sample drawn from a zero-mean distribution, with standard deviation $\sigma$. Such a series of observed data points may be obtained, say, from the calibration of a three-dimensional light stripe sensor mounted on a robot arm. The goal is to select the polynomial model which most effectively describes the data. The order (degree) and the parameters (coefficients) of the polynomial is to be determined by the algorithms. The simplicity of the model is important, since it would be used in real-time processing of light stripe sensor data.

A least squares error (LSE) estimation method is used to find the polynomial parameters for orders $D = 0, \ldots, 10$. As can be seen from Figure 1.3, for $D = 0$ the model is the simplest but the estimation errors are the large, while for $D = 10$ the estimation errors are reduced to zero by

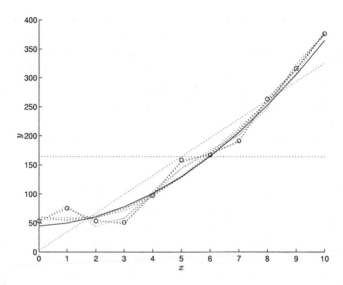

Fig. 1.3  Polynomial model order selection and parameter estimation. The circles indicate data points generated by adding zero mean Gaussian noise to the second order
polynomial shown by the solid line. The dotted lines show the polynomials of orders
$D = 0, \ldots, 10$, obtained from a least squares fit to the noisy data points.


using a more complex model. Figure 1.4 shows the representation size for
this problem plotted against the order of the polynomial. The minimal
representation size model, $D = 2$, is chosen as the best tradeoff between
the model size and the error residuals. Indeed, this was also the order of
the underlying polynomial model used to generate the noisy data.

For completeness' sake, the representation size of the observed data set,
$\mathbf{Z}$, shown on this plot, is given by:

$$\mathcal{L}[\mathbf{q}_D, \mathbf{Z} \,|\, \mathbf{Q}] = \underbrace{\sum_{d=0}^{D} c_d}_{\text{Model Size}} + \underbrace{\sum_{i=1}^{N} \log_2(\frac{|\mathbf{z}_i - \tilde{\mathbf{z}}_i|}{\sigma} + 1)}_{\text{LSE Error Representation Size}} \qquad (1.1)$$

where $\mathbf{q}_D$ is the $D$-th order polynomial, whose parameters $\boldsymbol{\theta}$ are obtained
from the LSE estimation, and $\tilde{\mathbf{z}}_i = \mathbf{q}_D(x_i) = \sum_{d=0}^{D} \theta_d x_i^d$ is the *predicted*
data point using this estimated polynomial. The set $\mathbf{Q} = \{\mathbf{Q}_0, \ldots, \mathbf{Q}_{10}\}$
is a library of polynomial models under consideration, where $\mathbf{Q}_D$ is the set
of all polynomials of degree $D$. The model representation size is obtained
by encoding each parameter $\theta_d$ in $c_d$ bits. In general, $c_d$ is also selected

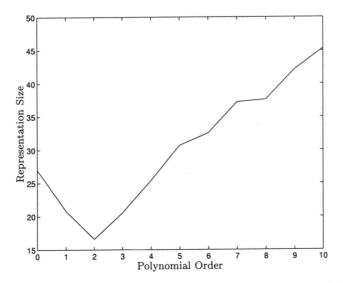

Fig. 1.4 Representation size vs. the order of the polynomial for the curve fitting problem. The minimal representation size model, $D = 2$, is chosen as the best tradeoff between the model size and the error residuals.

automatically by the minimal representation size criterion; it increases as the number of data points $N$ increases, and as the noise $\sigma^2$ decreases (i.e., as data accuracy improves).

This simple example helps to provide intuition about the role which the representation size plays as an information measure in choosing a simpler model than that which purely minimizes fitting errors (e.g. $D = 10$). This same principle of simpler models combined with consistency of correspondence and implicit scaling among disparate sensors is especially effective for the multisensor fusion problem.

## 1.4  Main Contributions

The main contributions of this book are summarized below.

- A broadly applicable framework for multisensor fusion and model selection, which addresses all the issues raised in Section 1.2. The theoretical framework is characterized by:

- General constraint equations that relate observed data features to environment model features.
- Various uncertainty models for observation errors, including probabilistic, and the newly developed encoded accuracy and precision based uncertainty models.
- Several correspondence types, which describe different constraints on mappings from data to model features.
- Library of environment models, which may vary in structure and the number/type of parameters.

The framework uses the minimal representation size approach, and prescribes a methodology for formulating a minimization criterion, for a variety of multisensor fusion and model selection problems.

- Search algorithms for multisensor data fusion and model selection.

  - Polynomial time fusion algorithms using the heuristic of constraining data feature sets (CDFS) to hypothesize environment model parameters, and using graph matching algorithms to find the minimal representation correspondences.
  - Evolution programs which are often quite practical, but usually offer no analytical bounds on convergence rates or guarantee of finding the global minimum.

The fusion algorithms simultaneously estimate the environment model structure, parameters, and the data correspondences, which describe the "best" interpretation of the observed data.

- Application of the multisensor fusion framework for two and three-dimensional object recognition and pose estimation using multisensor data.

  - Development of a non-redundant representation for three-dimensional rotations, which uniformly samples the space of rotations and is well suited for evolutionary search algorithms.
  - Laboratory demonstration and verification of the method for the tactile–visual fusion problem shown in Figure 1.1 on page 2.

- A highly reusable object-oriented application framework called MOMFIS (**MO**del-based **M**ultisensor **F**usion and **I**nterpretation **S**ystem), implemented in the *C*++ programming language. MOM-

FIS takes a modular component based approach to software development, and the software application framework classes embody the mathematical multisensor fusion framework described in this book. To apply the MOMFIS application framework on a new problem domain, the programmer has to derive a few classes and "fill in the blanks", for that specific problem domain.

## 1.5 Book Outline

In Part 1, we present the background material for this book. In Chapters 1.5 and 2.6.2 we review the previous work on multisensor fusion and integration. Chapter 4 describes the previous work on object recognition, localization, and matching. In Chapter 5 we provide an introductory tutorial on minimal representation principles, and introduce the notation and encoding schemes for fundamental entity types.

In Part 2, we present the minimal representation size framework for multisensor fusion and model selection. In Chapter 6, we introduce the different types of models used in the framework, develop an abstract problem description, and introduce the associated notation. In Chapter 7, we develop the minimal representation size framework, which prescribes a methodology for applying minimal representation size principles to a class of multisensor fusion and model selection problems. We develop encoding schemes for different choices of uncertainty models and correspondence types, which result in various multisensor fusion criteria, each suited to the specific problem type. In Chapter 8, we present search algorithms for minimizing the representation size criterion. These include a polynomial time hypothesize and test algorithm, and evolutionary algorithms.

In Part 3, we apply the framework to solve multisensor object recognition, pose estimation, and matching problems. In Chapter 9 we outline a recipe for applying the abstract framework developed in Part 2, to concrete problems. In Chapter 10, we illustrate the framework on two-dimensional problems, where the object identity, scale, position, orientation, and data correspondences are to be estimated from observed tactile and vision data; simulation results are presented. In Chapter 11, we apply the framework on three-dimensional problems, where the object identity, position, orientation, and data correspondences are to be estimated from tactile, vision, and grasp data; simulation results are presented.

In Chapter 12, we present the experimental results from our elaborate laboratory setup, for the vision–tactile fusion problem introduced in Figure 1.1 on page 2. In Chapter 13, we discuss the experimental results, and use them to illustrate the properties of the minimal representation size multisensor fusion framework. In Chapter 14, we summarize the work and point out a few open areas for research.

The symbols used in this book are summarized in Appendix A. The MOMFIS software architecture is described in Appendix B.

Portions of the work described in this book have appeared in the published literature [95, 97–104].

# PART 1

# Background

Chapter 2

# Introduction to Multisensor Integration

## 2.1 Multisensor Integration

*Multisensor integration* refers to the use of multiple sensors in a system; and usually suggests a *synergistic* use of multiple sensors, leading to improved operation of the system as a whole. Multisensor integration may require knowledge of the task to be accomplished, and the representation of that knowledge is often a key challenge in system design.

The literature on multisensor integration can be divided into three broad categories: **(a)** multisensor data fusion, **(b)** multisensor planning, and **(c)** multisensor system architecture, as shown in Figure 2.1. *Multisensor data fusion* deals with the *combination* of data from multiple sensors into one

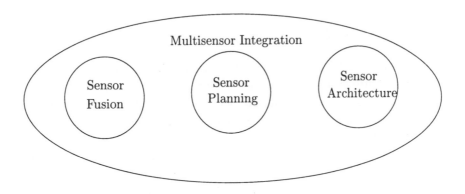

Fig. 2.1   Multisensor Integration.

coherent and consistent internal representation or action. *Multisensor planning* deals with the *acquisition* of sensor data. This may involve deciding which sensor to use next and deciding where to place it to get the best information. *Multisensor architecture* deals with the organization of data processing and *data flow* in the system.

Multisensor integration is an emerging discipline. In the literature, the term "multisensor integration" is often used to refer to any one or two or all of the above three categories (Figure 2.1 on the preceding page).

A broad tutorial on multisensor integration can be found in Hall and Llinas [76], which appears in the special issue on data fusion, of the Proceedings of the IEEE [157]. Another general review article can be found in Grossmann [73]. There are several review articles on multisensor integration for specific application domains: see Luo and Kay [123, 124] for intelligent systems, Kam et al. [106] for mobile robots, Leonard and Smith [119] for marine robotics, and Harris et al. [78] for multisensor data fusion in defense and aerospace.

Several books on multisensor data fusion and integration have recently appeared in the literature [2, 26, 27, 38, 125].

In the rest of this chapter we present a short overview of each of the broad multisensor integration categories described in Figure 2.1. We discuss sensor data fusion in Section 2.2, sensor planning in Section 2.3, and sensor architecture in Section 2.4. We discuss typical multisensor integration application domains in Section 2.5, and the current hardware and software technologies for multisensor integration in Section 2.6.

## 2.2  Multisensor Data Fusion

Multisensor data fusion deals with the fusion or *combination* of sensory data from multiple sensors into a *common representational format.*

A related, but more general, problem is that of *information fusion.* It refers to the combination of information not only from multiple sensors, but also from knowledge bases, databases, information processing blocks etc. into one representational format [76]. In other words, multisensor fusion is the version of information fusion restricted to combination of information (data) from sensor measurements. Information fusion includes multisensor fusion, decision fusion (combining expert "opinions" about a domain of discourse), and the combination of other information oracles.

We shall discuss multisensor data fusion, the main subject of this book, in Chapter 3.

## 2.3  Multisensor Planning

Multisensor planning deals with the acquisition of sensor data—such as deciding what data to acquire and how to acquire it. It usually includes the automatic formulation of sensing strategies to accomplish a task. This may involve deciding which sensor to use next and where to place it, and may use knowledge of the available sensors, actuators, the internal system state, the external environment state, and the overall task requirements.

The goal may be to minimize the time taken, efficiently utilize system resources, and/or minimize the number of sensing operations needed, while meeting "real-time" task deadlines. Often sensors and sensory data processing algorithms have time constants that vary considerably, and the sensing strategy must plan appropriately. Sometimes, *cooperation* of sensors is necessary to accomplish a sensing task—one sensor may provide cues to guide the operation of another.

Hager [74] divides sensing actions into two kinds: *performatory* actions that perform some subtask in order to get closer to the overall task, and *informatory* actions that gather more sensory data to enhance the knowledge of the external state and thus help in accomplishing the overall task. Sensor planning deals with deciding upon or planning the informatory actions to accomplish an overall task. It chooses the most appropriate configuration of sensors.

Sensor planning , also referred to as *sensor management* deals with managing a configuration of sensors, and choosing the best sensing configuration or actions. Sensor planning may be static or dynamic. *Static sensor planning* or *preselection* takes place during the system design, or during system initialization, and the sensor configuration does not change during system operation. *Dynamic sensor planning*, on the other hand, chooses the sensor configuration in response to changing environmental or system conditions during on-line system operation. It may be thought of as on-line planning of the "next-view", to accomplish a goal, given the current internal and external state.

Methods of static sensor planning include the work of Hager [74] and CAD-based vision techniques. Methods of dynamic sensor planning include the work of Hutchinson and Kak [84] for object recognition. They use Dempster-Shafer reasoning to maintain a set of object hypothesis about the environment, and choose subsequent sensors so as to maximally disambiguate between them. As a new datum is acquired a belief is assigned to it and used in updating the total belief in the object hypothesis. Ellis [54] describes a method to plan the best path for tactile probes in order to acquire data for object recognition. Manyika and Durrant-Whyte [132] describe a decision theory based approach to sensor management. Luo et al. [127] describe a multiagent based multisensor resource management system.

## 2.4  Multisensor Architecture

Multisensor architecture deals with the control organization and data flow in a multisensor system, to ensure that maximum benefit is derived from the use of multiple sensors. The focus is on the system aspects such as modularization, scheduling, coordination, robustness, and data communication among distributed measurement devices.

The multisensor system architecture should enable complementary sensors with non-overlapping ranges to operate together as a single sensor with greater reliability and robustness. Sensors may fail or violate the operating ranges. The architecture should degrade/upgrade gracefully as the set of operating conditions changes dynamically. The concern is with the efficient and proper flow of data.

Multisensor architecture may be divided into *data organization* which involves the representation, organization, and flow of the data within a system, and *control organization* which involves the control flow within a system. Typical sensing architectures for target tracking, identity fusion, high-level blackboard fusion reviewed in Hall and Llinas [76].

Some of the paradigms for data organization include logical sensors proposed by Henderson and Shilcrat [80], hierarchical phase-templates proposed by Luo et al. [126], artificial neural nets, and object oriented programming. Popular control structures include the RCS sensory and control hierarchy [11], distributed blackboards, and adaptive learning. Dasarathy [45] describes self-improving multisensor fusion system architectures using I/O based fusion modes. Other approaches to sensor organization, such as Bayesian nets and rule based systems, embed the data representation and the control flow in a single unified framework.

*Decentralized* sensing architectures use network of sensor nodes with local processing, which do not require any centralized facility for communication; only node to node communication and local system knowledge is permitted [132, 141]. *Distributed sensor networks* [89] use a network sensors to create a fault-tolerant sensing environment.

A formal semantics for a theoretical framework for an integrated software architecture for modeling sensor-based control systems has been developed by Dekhil and Henderson [46]. They define the components of a sensor system in terms of their functionality, accuracy, robustness, and efficiency, and develop a specification language to define sensor systems as a composition of smaller predefined components. Stieber et al. [186] present another systematic approach for the design of instrumentation architecture to enable the sensor fusion and robust control of complex electromechanical systems.

## 2.5   Applications

Multisensor integration has been employed in a variety of areas including robotics, guidance and control of autonomous vehicles, automated target recognition, target tracking, airborne surveillance, battlefield surveillance, pattern recognition, alarm analysis, monitoring of complex machinery, smart buildings, medicine and finance. Here are some of the application areas commonly reported in the literature.

**Mobile Robots** Multisensor fusion and integration is widely used in mobile robot applications and environments. Mobile robots typically employ a variety of sensors, including vision, range, sonar, infrared, and proximity [55], and are an active area of research on multisensor integration. Reviews of multisensor integration for mobile robots can be found in [10, 106, 131, 134].

Multisensor fusion techniques specific to mobile robots have been developed. Nygards and Wernersson [150] describe the use of covariances for fusing laser rangers and vision with sensors on-board a moving robot. Hexmoor [81] describe the fusion of vision, sonar, and contact sensory modalities in perceiving obstacles and targets; the arbitration of sensing and acting at reactive and deliberative levels; and integration of asynchronous instruction taking and communication.

**Object Recognition** Object recognition seeks to automatically identify
objects and their parameters using a variety of sensors, such as vision, range, laser, touch, and so on, in structured, semi-structured,
and unstructured environments. This is another active area of multisensor fusion research, and is discussed in Chapter 4.

**Target Tracking** Target tracking, motivated by use in military and defense, seeks to combine data from multiple sensors observing multiple targets, and automatically track several moving targets over
time. It is another active area of multisensor fusion and integration
research [86, 196, 202].

**Medical** Recently multisensor fusion is being used in medicine to combine multisensor sensor measurements, and deduce a more accurate assessment of the state of the human body, and to diagnose
ailments. Thoraval et al. [195] describe the use of multisensor fusion
to fuse electro-physiological and haemodynamic signals for ventricular rhythm tracking.

**Testing** Multisensor fusion is being applied to non-destructive testing,
where data from multiple sensors is fused to deduce hidden de-
fects and faults. Gros [72] describes the use of Dempster-Schafer
fusion of Gaussian densities to improve overall efficiency of non-
destructive examination, and shows that the probability of error
decreases as the number of sensors increases.

**Miscellaneous** Multisensor fusion has been applied to a variety of other domains [171]. Wide et al. [209] describe a multi-sensor fusion
method for artificial nose and tongue sensor data for chemical, pa-
per pulp, food, and medicine applications. Nelson and Fitzgerald
[148] describe the use of sensor fusion for alarm analysis, by fusing
visible, infrared, and millimeter wave radar data.

## 2.6    Technologies

Advances in multisensor fusion and integration go hand-in-hand with the
development of new sensor technologies, and the availability of tools to
utilize the sensing technologies.

### 2.6.1  *Sensors*

New sensor technologies are constantly being developed, based on inno-vative applications of physical principles to sensing problems. Everett [55] gives a comprehensive guide to sensors developed in the context of mobile robots. Recently there has been much interest in *microsystems*—microscopic machines and sensors on a chip. Luo [128] give a good overview of sensor technologies and microsensor issues for mechatronics systems. The industry has been moving towards the use of smart sensors, fieldbus net-works, and localized smart processes to standardize the integration and use of multisensor technologies [193].

### 2.6.2  *Tools*

A big challenge in building multisensor systems lies in addressing the soft-ware complexity inherent in integrating multiple sensors and other parts of a system. The availability of software tools can make development of mul-tisensor systems easier, and facilitate greater experimentation with new algorithms and sensors.

The notion of "Instrumented Logical Sensor Systems" developed by Dekhil and Henderson [46] includes set of tools for dynamic performance analysis, on-line monitoring, and embedded testing of sensors systems. Hal-l and Kasmala [75] have developed a visual programming environment to support the implementation of various data fusion algorithms and facili-tate rapid prototyping of data fusion systems. *ControlShell*\* [161, 174] is a commercially available component-based programming system specifically tailored to intelligent control applications, which facilitates software inte-gration by hierarchical composition of well-defined simple generic primitive components into complex problem specific components.

To facilitate experimentation and the simplify the application of the multisensor fusion methodology developed in this book, we have develope-d a flexible and extensible software tool called MOMFIS (**MO**del-based **M**ultisensor **F**usion and **I**nterpretation **S**ystem) [95]. It includes a *C++* class library, and to address a new application domain, the programmer has to derive a few classes and "fill in the blanks", to define the new appli-cation domain.

---

\*The first author is a developer of *ControlShell*

# Chapter 3

# Multisensor Data Fusion

## 3.1 Introduction

Multisensor data fusion refers to the combination of sensory data from multiple sensors into a *common representational format* [73, 76, 112, 123, 163] (Figure 3.1). The focus is on *combination* of data from various sensors into a single inference or interpretation, without regard to how it is obtained, or how the sensors are organized. We assume that an appropriate sensing architecture and sensing plan has been devised to collect data.

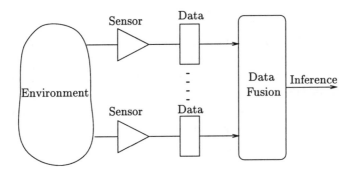

Fig. 3.1  Multisensor Data Fusion.

### 3.1.1 *Why Multisensor Data Fusion?*

Multisensor data fusion is employed in applications for a number of reasons. The most commonly cited reasons are listed below.

(1) **Accuracy:** The inferences made based on a set of sensors are likely to be more accurate, than those based on a single sensor alone. Redundant information from *competitive* sensors, about the same characteristics of the environment can be used to reduce the uncertainty in the fused inference.

(2) **Synergy:** A single sensor's data may be incomplete, but *complementary* sensors observing different aspects of the environment can be used to *synthesize* inferences that are impossible to make based on an individual sensor alone.

(3) **Reliability:** Use of multiple sensors increases *fault tolerance*, and improves system reliability—even if a few sensors fail, the fused inference is still likely to be accurate. If a sensor fails, another sensor can provide similar information.

(4) **Robustness:** Multiple sensors (that may be based on different physical principles) can be used to increase the effective range of situations in which the system can operate, making it robust to large fluctuations in "operating conditions".

(5) **Timeliness:** The inferences made with multiple sensors can be more timely than those obtained from a single sensor due to faster sensors, overlapping time scales, or due to the processing parallelism (depending upon the modularity of the multisensor fusion method).

(6) **Lower Cost:** It may be less expensive (in terms of processing time, computational resources, or hardware resources) to obtain the same inference by fusing multisensor data, than with individual sensors, provided the cost of sensors themselves is disregarded*. Multisensor fusion may save the cost of adding additional sensors; conversely adding more sensors may save on the processing cost.

---

*The cost of sensors can be amortized over the system life-span, and may be negligible compared to the operating cost.

### 3.1.2 *Problems and Issues*

In this book we deal with the question of *how* to fuse the observed multisensor data. A technique for multisensor data fusion should consider several key issues, listed below.

(1) **Registration:** Each sensor provides data in its local reference frame. The data from different sensors must be converted into a common reference frame before combination. This problem of *aligning* sensor reference frames is often referred to as the registration problem. Errors in sensor *calibration* must be accounted for at this stage.

(2) **Homogeneous vs. Heterogeneous data:** Different sensors may provide *qualitatively* the same (homogeneous) or different (heterogeneous) types of data. Fusion schemes ought to handle both homogeneous and heterogeneous data features.

(3) **Uncertainty in Sensor Data:** Sensor data is inherently uncertain. The uncertainty may arise due to sensor noise or due to spurious measurements (outliers). Consequently, sensors may provide *competitive* or *conflicting* data. Fusion schemes ought to exploit the redundancy in the data from competitive sensors to reduce the uncertainty in the fused inference, while rejecting conflicting data due to outliers.

(4) **Incomplete, Inconsistent, and Spurious Data:** When multiple interpretations can give rise to the same observed sensor data, we say that the data is *incomplete*. This is usually due to an insufficient number of data features obtained from the sensors. The observed data features do not contain enough "information" to determine a unique interpretation. The multisensor data set can be made complete by **(a)** collecting more data features, or **(b)** by increasing the number of sensors.

We say that two sensors are *inconsistent*, if their data sets are complete, but are described by different interpretations. In other words, several distinct interpretations give rise to the observed (complete) sensor data sets. This usually results due to **(a)** improper sensor registration (e.g. poor calibration), or **(b)** the sensors observing different environments.

When the sensor data contains features that cannot be related to the observed environment model, we say they are *spurious* data

features. When more than 50% of the data features are spurious,
they are impossible to distinguish from incomplete or inconsistent
data, on the basis of the single sensor data set alone. Other correctly registered sensors can be used to select a consistent subset
of data and discard the spurious features.
An effective multisensor fusion technique ought to automatically
synthesize correct interpretations by appropriately fusing incomplete, inconsistent, and spurious data.

(5) **Correspondence/Data Association:** Once the sensors are registered, we still need to establish which data features in one sensor
refer to the same aspect of the environment in the other sensor.
In other words, which data features in one sensor are associated
with which data features in the other sensor, and which data feature are outliers? This is the correspondence problem, or the data
association problem. If we have a model of the environment, it is
tantamount to establishing a correspondence between the data and
the model features for each sensor. Fusion schemes must address
the issue of handling data correspondences.

(6) **Sensor Dependencies:** Sensor data may not be independent.
Some of the sensors may be affected by a common external dis-
turbance, that may bias each sensor's data. It is important to
remove such data dependencies, otherwise the sensors (taken together) may be perceived as more/less accurate than they actually
are. *Cooperative* sensors may form a chain, such that the output
of one sensor depends on that of the previous one. Fusion schemes
ought to address the issue of handling such data dependencies.

(7) **Granularity:** Different sensors may provide data at different levels of granularity. The data may be *sparse* or it may be *dense*.
Furthermore, data may be at different levels of abstraction (or
processing)—*signal, pixel, feature, and symbol* levels are frequently
encountered. A fusion schemes ought be able to work at various
granularity levels.

(8) **Static vs. Dynamic:** Sensors may be observing a static environment, or a dynamic time varying one. The past *history* may be
important for a dynamic environment, and there ought to be a way
of incorporating this in the fusion scheme.

(9) **Time Scales:** In a dynamic scenario, different sensors may have

different time constants and operate at different rates. Also, different aspects of the environment may be varying at different rates. A fusion methodology ought to incorporate multiple time scales and time-varying data.

(10) **Modularity:** It is desirable to independently process multisensor data, as much as possible. At some stage all the data processing will have to come together for the fusion to take place. A modular technique will make it possible to map the computation into parallel and/or distributed processors. In a distributed computing environment, such as ours, each node may locally process sensor data concurrently. The partially processed sensor data would be fused by a central fusion node, that gets its inputs from all the sensor nodes.

A single fusion scheme seldom addresses all the above issues in depth—and indeed the various techniques in the literature emphasize one or a subset of the above issues. In practice, a combination of techniques with different strengths may be used, to solve a specific multisensor data fusion problems.

### 3.1.3 *Forms of Multisensor Data Fusion*

There are two major sensor fusion paradigms in the context of intelligent systems. One paradigm (explicit representation) is to build an internal representation or "environment model" from the sensed data. A separate planning phase decides what action to take, based on the internal representation:

Sensing $\longrightarrow$ Representation $\longrightarrow$ Action.

The second paradigm (implicit representation) links sensing directly to action without requiring an explicit internal representation [116]:

Sensing $\longrightarrow$ Action.

Multisensor fusion can be studied in the context of either paradigm, and refers to the combination of data from multiple sensors into one common representation or action. In practice, a combination of the two paradigms may be necessary to solve real-world problems.

In *homogeneous* multisensor data fusion, data samples from a single type of sensor are combined to build an estimate or interpretation of the environment. In *heterogeneous* multisensor data fusion, data samples from several different types of sensors are combined to build an estimate or interpretation of the environment.

The common fused representation may range from a low-level probability distribution for statistical inference to high level logical propositions used in production rules for logical inference. Luo and Kay [124] divide the previous approaches to multisensor data fusion into four categories: *signal, pixel, feature,* and *symbol* level fusion. Each successive fusion level is more abstract and at a higher representational granularity. At these different levels, it is customary to assume that sensor measurement models are known, and the sensor data from different sensors is independent, with the observation errors modeled by statistically independent additive noise.

*Signal-level* fusion deals with the combination of signals from a group of similar sensors with the aim of deriving a single composite signal, usually of the same form as the original signals but higher quality. A high degree of spatial and temporal registration or data association between the signals is necessary for fusion to take place. Signal-level methods include weighted averaging, Kalman filtering, Bayesian estimation using consensus sensors [126], and statistical decision theory methods [135].

*Pixel-level* fusion deals with the combination of multiple images into a single image with a greater information content. A high degree of spatial correspondence or data association between the images is necessary for fusion to take place. Pixel-level methods include logical filters, mathematical morphology, image algebra, and simulated annealing.

*Feature-level* fusion deals the combination of features derived from signals and images into meaningful internal representations, or more reliable features. Feature-level methods include tie statistic, Gauss-Markov estimation with constraints, and extended Kalman filters. The methodology developed in this book falls under this category.

*Symbol-level* fusion deals with the combination of symbols with an associated uncertainty measure, each representing some decision, into symbols representing composite decisions. It is sometimes also referred to as *decision fusion.* Symbol-level methods include Bayesian estimation, Dempster-Shafer evidential reasoning, production rules with confidence factors, and fuzzy logic.

Multisensor fusion scenarios can be divided into three categories: complementary, competitive, and cooperative [51]. *Complementary* sensors observe different aspects of the environment, each providing partially overlapping incomplete data. This is the simplest scenario; the partial data from the sensors can be used to *synthesize* inferences that are impossible to make based on an individual sensor alone. *Competitive* sensors provide overlapping redundant data about the same aspect of the environment. The redundancy can be used to reduce the uncertainty in the fused inference. Outliers may lead to conflicting sensor data, and should be detected. *Cooperative* sensors guide one another, by exchanging sensory cues. They may form a chain where the output of one sensor depends on that of another; thus there is an inherent dependency between them. All the three scenarios may manifest at the same time in a real system, and often may not be separable.

## 3.2  Modeling in Multisensor Data Fusion

Multisensor data fusion involves the use or construction of models from multisensor data.

### 3.2.1  *Environment Modeling*

Environment modeling refers to the representation of the *subject* under observation by various sensors. The environment models may be explicit or implicit, and embody the common representation format among multiple sensors. Explicit models usually include parametric models [19], nonparametric models, geometric models and so on. Implicit models include rule-based models, neural network representations, and so on [201]. We discuss generic environment models in Section 6.1.

A big challenge in multisensor fusion is *model selection*—selecting an appropriate environment model based on observed multisensor data. Many traditional approaches assume that the model structure is known, and focus on estimating its parameters. In this book, we develop a multisensor fusion framework, which addresses both the model structure selection, and the parameter estimation problems simultaneously (Chapter 7).

### 3.2.2  *Sensor Modeling*

Sensor modeling refers to the use and construction of sensor and observed data models. It includes (1) the *physical model* which relates the observed data to the subject under observation, (2) the *uncertainty model* which describes the random errors (noise) introduced in the observation process, and (3) the *correspondence model* which describes the association of observed data features to features in the subject under observation, and also describes which data features are spurious. We discuss generic sensor models in Chapter 6.

Often the physical sensor models are well-known from the sensor Physics, and one one must devise an appropriate sensor uncertainty model. Various approaches for uncertainty modeling have been suggested in the literature, including probabilistic models, fuzzy models, belief models, geometric model and so on. An overview of some of these approaches can be found in Lehner et al. [118], which appears in the special issue on higher order uncertainty, of the IEEE Transactions on Systems, Man and Cybernetics [181].

Another big challenge in multisensor fusion is *correspondence model selection*—selecting an appropriate correspondence based on observed multisensor data, given the physical sensor model, and the sensor uncertainty model. Many researchers have focused on the difficult problem of filtering spurious data features (or outliers) [109]. The multisensor fusion framework developed in this book (Chapter 7), is capable of automatically selecting the correspondence model from observed data.

## 3.3   Multisensor Data Fusion Techniques

The goal of multisensor fusion is to combine data from multiple sensors to do any or all of the following (1) find the environment model structure, (2) find the environment model parameters, (3) find the sensor data correspondence, or (4) make a decision, or (5) take an action. The techniques to achieve these goals have been taken from many diverse areas, including artificial intelligence, pattern recognition, statistical estimation, and so on.

Sometimes the data fusion techniques are tailored to an application domain, or are specific to the type of sensors used. We shall discuss techniques specific to the object recognition application domain in Chapter 4.

Here we describe some of the generic multisensor data fusion techniques. Other reviews may be found in Abidi and Gonzalez [2], Brooks and Iyengar [26], Crowley and Demazeau [42], Hall and Llinas [76], Kokar and Kim [112], Luo and Kay [124].

### 3.3.1  *Signal Processing Methods*

Signal processing methods are typically used for low-level signal fusion [42]. These include filtering and signal enhancement techniques. Chipman et al. [36] describe the use of wavelets to fuse images from different wavelengths. Some widely used techniques are described below.

#### 3.3.1.1  *Weighted Averaging*

A common approach for homogeneous sensors is to take the weighted average of the various sensor data values or their interpretations, to arrive at a composite fused value. All the input data must be of similar type. The fused value may be used to directly trigger an action. In a heuristic approach, the weights are fine-tuned until the system has a desirable performance; in a more sophisticated approach, such as Kalman filtering, the weights can be chosen to meet some performance criteria. This method is applicable for dynamic and redundant data that is changing in time, and is a signal-level fusion technique [124].

#### 3.3.1.2  *Kalman Filtering*

Kalman filtering is a popular method for state estimation that recursively updates its estimate of the probability density function defined on the environment state, as new sensor data is acquired; it may use a *known* system model to predict the state transition from one sensor datum to the next. Kalman filtering can be regarded as a signal-level fusion technique.

Kalman filtering is a linear systems technique that works well for reconstructing the environment, when the data is corrupted by measurement noise only. It is different from conventional filtering methods in that it uses an explicit (probabilistic) system model. The system model is represented by a state vector to be estimated, a known state transition matrix, and an additive zero mean white noise process with a known covariance matrix.

A linear measurement model is assumed—the measured data is a linear transformation of the unknown state vector corrupted by zero mean white measurement noise. A Kalman filter builds an unbiased, minimum variance, consistent estimate of the system state vector by optimally combining its predicted estimate from the system model (using the state transition matrix) and the estimate from the measured data. It leads to recursive equations for state vector update, as more data is acquired. The extended

Kalman filter (EKF) generalizes the technique to nonlinear systems by linearizing the measurement and system models. Kalman filtering has been used for localization [39, 44, 141] in mobile robot navigation.

The Kalman filtering approach assumes that the structure of the system model is known *a priori*. This approach is useful when the state vector can be identified and related to its previous values through a state transition matrix. The noise models for the system model and measurement model are very simple—additive zero mean white noise processes. Such an approach is however very sensitive to outliers in the data; they can completely throw off the estimate of the system state vector. Other techniques must be used to filter the outliers from the data.

### 3.3.2   Estimation Theory Methods

#### 3.3.2.1   Least Squares Estimation

Least squares estimation methods [184] fuse data by searching for solutions which minimize the squared error between the observed data and the predicted data.

Shekhar et al. [178] describe a weighted least squares method to compute the position and orientation parameters of an object from sparse contact point tactile data. It is assumed that object identity and the correspondence between the object model and the data is known. First, a set of vectors is computed from the measured points, and from it a subset of best vector features is chosen to minimize the expected error in the orientation. Next, the orientation parameters (represented by quaternions) are expressed as a linear system of the best vectors, and their optimal estimate is obtained by solving a weighted least squares linear system of equations involving the best vectors. The position parameters are computed similarly. The weights are assigned inversely proportional to the error in the estimate of the vector features for orientation computation, and to the error in measured point features for position computation. Outliers in the data can bias the pose estimate significantly; this issue is not addressed.

Eason and Gonzalez [52] use a least squares approach to fuse multisensor data to determine the pose of a known object. The correspondence between the model and the data features is assumed to be known. The data obtained from the various sensors is regarded as a set of constraints on the pose transform space. The measured sensor data along with its distribution is used to provide a measure of error on the transformed model features, as a function of the object pose. The pose transform that minimizes total error over all the features is chosen to be the best solution. As more data is collected this solution space is pruned until a final solution is found. The sensor data could be coming from vision, force/torque (touch), proximity, and ultrasonic range sensors from which point, line, or planar face features are obtained.

### 3.3.2.2 *Risk Minimization*

Richardson and Marsh [163] discuss a state estimation approach based on minimizing the expected loss or risk. The loss function is defined as a function of the actual value and the estimated value of the state. The problem is reduced to that of finding the optimal functional that expresses the state estimate as function of the observed data. They also outline a pseudo proof that the inclusion of additional sensors almost always improves the performance of any fusion scheme based on optimal estimation. They also discuss the modularization problem, and describe an example of fusing acoustical data with laser data. This method may be considered to be a signal-level fusion technique.

### 3.3.2.3 *Energy Minimization*

Energy based methods define an energy/cost function to be optimized, to find a fused representation. The energy function may be defined heuristically, or may be defined to meet some performance criteria, and is the subject of much study.

Energy based methods for data fusion minimize some energy function of the data obtained by different sensors and a prior model of the phenomenon under observation. The prior model is often represented as constraints on the observed data. Regularization terms may be introduced that minimize some internal energy [107, 194, 214, 218]. Clark and Yuille [38] describe a general description of these methods, and point out their relationship to the Gibbs distribution and Markov Random Fields [64]. The methodology developed in this book leads to an optimality criterion to be minimized, and may be viewed as an energy minimization approach.

Kass et al. [107] have proposed "snakes" an approach to image reconstruction with prior model information represented by a regularization term. A *snake* is an energy minimizing spline that is guided by external constraint forces and influenced by image forces that pull it towards features of interest. The smoothness constraint or the regularization term is introduced as the internal energy of the spline. The external constraint forces may be cues provided by the high level processing of what to expect. The image forces are the significant features in the data to which the spline is attracted. Minimizing the total energy due to the internal and external constraint and image forces, produces a spline contour as the solution.

Yamada et al. [212] use an energy function minimization approach to reconstruct shape represented by deformable superquadric models, from visual and tactile data. A global superquadric model is first fitted to range data obtained from a visual sensor. Portions of this recovered superquadric model that are invisible or have large fitting error are then explored by a tactile sensor, and deformed locally. The local deformations are applied so as to minimize an energy function comprised of an internal energy term and an energy term based on the fit of the the tactile datum to the reconstructed surface.

### 3.3.3  *Statistical Inference Methods*

#### 3.3.3.1  *Maximum Likelihood Estimation*

Maximum likelihood (ML) methods [184] fuse data by searching for solutions which maximize the probability of observing the collected multisensor data.

Zhou et al. [216] describe a maximum likelihood method for multisensor data registration. Luo and Tsitsiklis [129] describe a decentralized fusion architecture using maximum-likelihood estimation, and exploiting the optimality of Gaussian random variables.

Bolle and Cooper [19] describe a Bayesian approach to estimation of the pose of an object from range data, given the object identity and the correspondence between the model and the data features. They assume that the segmentation of the range data is given, and use Bayesian estimation to locally fit a plane, cylinder, sphere, line or a circular curve to each segment. The local maximum likelihood parameter estimates from each segment fit are combined in a second step to compute the global maximum likelihood object pose parameter estimate. Features are not computed explicitly, the data set is dense, spurious data are not modeled explicitly, and only simulation results are presented. Computing the global pose parameters from the local parameters reduces to minimizing a Mahalanobis distance function. They use a nonlinear optimization technique, which can get stuck in local minima and may not be suitable for real-time implementation. This approach may be regarded as feature-level fusion.

### 3.3.3.2 *Maximum A Posteriori Estimation*

Maximum a posteriori (MAP) methods [47, 184] fuse data by searching for solutions which maximize the probability of observing the collected multisensor data given a prior distribution on the possible solutions.

Durrant-Whyte [50, 51] describe a scheme for combining partial, uncertain geometric observations in a robotics setting into a consistent estimate of the environment state. The environment is regarded as a collection of uncertain geometric objects. Each object is represented by a $\epsilon$-contaminated Gaussian distribution on the parameters describing it. The sensors are regarded as a team of decision makers, each sensor being a Bayesian estimator. A joint *a posteriori* distribution on an object is constructed by combining the probability distributions on the object due to each sensor. The likelihood of this joint distribution is maximized. The fused information together with a prior environment model can be used to direct the robot system during the execution of different tasks. The basic idea of combining the pdfs is similar to that in Kalman filtering. This work does not address the data association or the correspondence problem. It may be regarded as a signal-level fusion method.

Wells III [208] uses a maximum a posteriori (MAP) estimation criterion to match an object model to the data for the two-dimensional object identification and pose estimation problem. The data features matched to the model features are assumed to be described by a Gaussian distribution. The background data features or outliers are accounted for explicitly by a uniform probability distribution. The prior on the pose parameters is also assumed to be Gaussian. A prior is also assumed for the correspondence. The MAP technique is used to obtain the estimates for the pose and the correspondence that maximize the a posteriori probability. Heuristic beam search and least squares optimization find solutions (pose and correspondence) that maximize the objective function.

Chen and Ansari [35] describe an optimal decision rule, based on a the maximum posterior probability detection criterion, for adaptive fusion of correlated local decisions.

### 3.3.4   *Information Theory Methods*

Information-theoretic methods [17, 40] fuse data by searching for solutions
which optimize various information measures in a multisensor system. Basir
and Shen [14] utilize an *information variation* measure for modeling uncertainty and cooperation among sensors, and develop a stochastic weighting
scheme, which operates recursively on a sensor team until it reaches a consensus. Mutambara [141] describes scalable decentralized estimation and
control algorithms for linear and non-linear multisensor systems, using a
decentralized extended *information filter*. Some widely used information-
theoretic techniques are described below.

#### 3.3.4.1   *Entropy Based Methods*

Entropy based methods methods fuse data by searching for solutions which
minimize the uncertainty in a multisensor system. Zhou and Leung [217]
describe a minimum entropy approach for multisensor data fusion in non-
Gaussian environments. They represent the fused data in the form of the
weighted sum of the multisensor outputs and use the varimax norm as the
information measure. The optimum weights are obtained by maximizing
the varimax norm of the fused data. The minimum entropy fusion solution
only depends on the empirical distribution of the sensor data and makes no
specific distribution assumptions about the sensor data. Chung and Shen
[37] develop a team consensus approach based on information entropy, for
fusing dependent sensory data. Manyika and Durrant-Whyte [132] describe
an entropy based approach for data fusion and sensor management.

#### 3.3.4.2   *Minimum Description Length Methods*

Minimum description length (MDL) or minimal representation size (MRS)
methods [12, 13, 104, 142, 169] fuse data by searching for solutions which
minimize the description length (representation size) of observed multisensor data. The description length or representation size, is defined as
the length of the shortest program which can reproduce the observed data [177]. The framework developed in this book uses representation size to
fuse multisensor data.

Foster [60], Sanderson and Foster [173] describe the use of minimal
representation size for two-dimensional pose estimation from image data.
Ravichandran [158], Ravichandran and Sanderson [159, 160] uses represen-

tation size for three-dimensional pose estimation from image data. We have used representation size for estimating three-dimensional pose from grasp data [97], and for fusing vision and touch data for two-dimensional [98], and three-dimensional [95, 100–104] object identification, pose estimation, and correspondence assignment.

Fua and Hanson [61] use the minimum description length criterion to set up an objective function to be maximized, for two-dimensional feature extraction from image data. This work illustrates the application of MDL to a data fitting problem. The objective function takes into account the representation cost and the quality of the shapes, and balances the geometric requirements against photometric evidence from the image data. The system uses underconstrained generic object models. Thus, the model may specify that a generic polygon is to be extracted from the data. Its shape and size is unknown and inferred from the data as a part of the generic model recovery process.

### 3.3.5   *Decision Theory Methods*

Decision theory methods [15, 68] are typically used for high-level decision fusion. Iyengar and Prasad [90] describe a general computational framework for distributed sensing and fault-tolerant sensor integration. Nelson and Fitzgerald [148] use decision theory methods to fuse visible, infrared, and millimeter wave radar for alarm analysis. Varshney [202] describe techniques for decision making in a distributed network, and utilize statistical decision theory.

### 3.3.6   *Artificial Intelligence Methods*

Many artificial intelligence methods [162] have been applied to or developed in conjunction with solving data interpretation problems. Some of these are described below.

#### 3.3.6.1   *Dempster-Shafer Theory*

Dempster-Shafer (DS) theory of evidence is based on the notion of assigning *beliefs* and *plausibilities* to the possible interpretations of observed multisensor data. Bloch [18] presents a survey and taxonomy of various belief fusion operators, and provides a guide for choosing an appropriate operator for belief combination. Gros [72] used Dempster-Shafer fusion of two Gaussian densities for non-destructive testing. Murphy [140] applies Dempster-Shafer theory to sensor fusion in autonomous mobile robots.

### 3.3.6.2  *Fuzzy Logic*

Fuzzy logic introduces the notion of a partial set membership, and has been used by several researchers for fusing multisensor data. Goodridge et al. [69] describe a modular fuzzy control architecture where the control function is broken down into multiple local agents, each of which samples a subset of a large sensor input space. They use fuzzy agents to fuse the recommendations of the local agents. Yager [211] describes a general approach to the fusion of imprecise information, utilizing fuzzy measures. Runkler [172] describes model based sensor fusion using a fuzzy model of the functional dependence between the sensor signals. The noisy sensor signals are fused by a projection onto the model, and lead to smaller errors.

### 3.3.6.3  *Rule-Based Reasoning and Expert Systems*

Symbolic reasoning and rule-based expert systems are typically used for symbol-level fusion. Rule-based classification methods map the multisensor data feature space to distinct labels or actions. Peers [154] describes a *blackboard* system that fuses data from electromagnetic and vision sensors, for automatic defect classification of automotive or other components.

### 3.3.6.4  *Neural Networks*

Neural networks have be used in data mining [210], classification [16], and in robotics to map the multisensor data space into actions for providing a real-time connection between sensing and action. Lee [116] describes sensor fusion and planning using a perception-action network, which consists of a number of heterogeneous computational units, representing feature transformation and decision-making for actions. These are interconnected together as a dynamic system. Input stimuli to the network invoke the evolution of network states to a new equilibrium, thus accomplishing real-time integration of sensing, knowledge, and action.

### 3.3.7  Geometric Methods

Geometric methods exploit the geometrical properties of the environment and sensor models to achieve multisensor fusion.

#### 3.3.7.1  Geometric State Fusion

Smith and Cheeseman [182] describe a method for estimating, combining, and propagating geometrical uncertainty (nominal relationship and expected error specified by covariance) associated with (two-dimensional) coordinate reference frames. A first order Taylor series approximation is used, and a Kalman filter is derived for the merging operation. Rules for compounding and merging approximate transformations (AT) are described, and their similarity to circuit theory is discussed.

Porill [155] describes a method for statistical combination of geometrical information from multiple sensors. A local reference frame is associated with each geometric primitive such as a point, line, or a plane. Small perturbations of the geometric primitive from this reference frame are described by an unconstrained representation in this local reference frame. The local reference frame coordinates of all the geometrical primitives together constitute a *state vector*, which is assumed to be a zero mean Gaussian process. Calibration parameters are also incorporated in this representation along with the measured sensor data. The fusion problem is is to find the best estimate of the state vector. Gauss-Markov estimation is used to derive a singular scalar Kalman filter, to update the estimate of the state vector, while satisfying the constraints imposed by the measurement model and other geometrical constraints imposed by the object model. The correspondence or the data association is assumed to be known. This technique could be regarded as feature-level fusion.

Intaek and Vachtsevanos [87] describe a geometric approach utilizing polygonal approximation and a matching task between the sensor data and stored templates, to fuse data from a suite of complementary sensors, for recognizing and identifying overlapping or occluded objects.

#### 3.3.7.2  Minimizing Uncertainty Ellipsoids

Nakamura [143] describes a geometric approach to low-level fusion of multisensor data, based on the minimization of the uncertainty ellipsoid volumes. The functional relationship (computational nonlinearity) between the measured sensor data and the high level sensory information derived from is is assumed to be known. This analysis methodology is based on first order Taylor series expansion and is similar to that used by Smith and Cheeseman [182], but is more generally applicable. This method reduces the amount of uncertainty in the fused value, which is computed as a weighted sum of the sensory information from different sensors. Structural (i.e. in the data generation mechanism) and computational nonlinearities are handled uniformly. The technique is shown to be equivalent to Bayesian inference using minimum variance estimate, Kalman filtering, and weighted least squares estimation. The method is extended to fuse partial information.

# Chapter 4

# Multisensor Fusion in Object Recognition

## 4.1 Introduction

Object recognition deals with the identification and pose estimation (the position and orientation in three-dimensional space) of geometric objects. The object recognition problem arises in semi-structured industrial automation environments, where information about the identity and pose of the objects in the workspace is needed. Commonly occurring tasks needing this information include materials handling, inspection, and assembly. The sensors used typically include workspace mounted cameras. Other kinds of sensors, such as "touch", structured light, and ultrasonic sensors, have been explored to facilitate and enhance object recognition. Often it is assumed that a library of models of all the objects that can occur in the workspace is available to the recognition system. The model library usually contains geometric descriptions of objects or object models, and may be derived form an industrial CAD database.

The object recognition problem may be broken down into further subproblems. The first is the problem of object *identification* which involves identifying or labeling an object in the workspace. The second is the problem of *pose estimation* which is to find the position and orientation of the object in the workspace, given its identity or class label. The third is the problem of finding the *correspondence*, which involves establishing the mapping or the association between the observed data features and the workspace object features.

Although, we use the term *object recognition* to refer to solving all the three problems above, it is often used to refer to just the object identifica-

Table 4.1   Forms of object recognition.

| Dimensionality of | | Number of Objects in | |
| --- | --- | --- | --- |
| Data | Model | Library | Workspace |
| $n_1, n_2, \ldots$ | $m$ | $L$ | $l$ |
| 2D or 3D, 2D or 3D, ... | 2D or 3D | Many (M) or One (1) | Many (M) or One (1) |

tion problem. The term *object localization* is used to refer to the problem of finding the pose and correspondence of a known object in the workspace. The emphasis is on computing the pose of the object, the correspondence is usually assumed to be known. The term *object matching* is used to refer to the problem of finding the correspondence and the pose of a known object in the workspace. The emphasis is on computing the correspondence, and the pose is often assumed to be known.

The model-based object recognition problem essentially has the flavor of a *data fitting* problem. Given a library of available object models, find the model that best fits the observed data. Note that the word *match* is often used in two contexts: in the sense of finding a correspondence, i.e., a match or a pairing of a model and data feature, and in the sense of fitting a stored object model to the data.

### 4.1.1   *Forms of Object Recognition*

The various forms of object recognition can be classified according to the type and number of of sensors, the dimensionality of the observed data from each sensor, the dimensionality of the object models, the number of object models in the library, and the number of objects under observation in the workspace, as shown in Table 4.1. In the 2D/2D/*/* form, both data and the objects are two-dimensional. Examples include an overhead camera observing a two-dimensional XY-table. In the 2D/3D/*/* form, the data is two-dimensional but the workspace and the objects are three-dimensional. Examples include a fixed vision camera observing a three-dimensional workspace. In the 3D/3D form, both the data and the objects are three-dimensional. Examples include a three-dimensional laser range sensor, or a coordinate measuring machine observing a three-dimensional workspace.

The difficulty of object recognition problems increases as the dimensionality of the data or the models increases, and as the number of objects on the library or in the workspace increase. The */*/1/1 form is the simplest with only one known object in the data, and its pose and the correspondence are to be estimated. The */*/M/M form is the hardest and the most general, with multiple (unknown) number of objects in the data, and their identities and pose are to be found. Much of the literature deals the */*/1/1 or the */*/M/1 forms of object recognition.

The literature on object recognition can be divided into *feature based* approaches, which process the sensor data to derive features from it; and *non-feature based* approaches, such as volumetric methods, which not compute explicit features from the observed data. Feature based approaches to object recognition may differ along a number of lines, including the sensors used to obtain the data and the number of sensors, the type of features extracted from sensed data, the uncertainty model for the data features, the object model (shape) representation, the model parameters to be estimated, the class of allowable transformations on the model, the method of estimating the model parameters (computing a pose transformation), the hypothesis generation strategy, the method used to establish the correspondence or match between the model and the data features, the criterion for verifying the correctness of the hypothesized model parameters (pose transformation).

We apply the minimal representation size multisensor fusion framework developed in this book to two-dimensional feature-based object recognition using vision and touch sensors (see Chapter 10), which is of the form 2D,2D/2D/M/1; and to three-dimensional feature-based object recognition using vision, touch, and grasp sensors (see Chapter 11), which is of the form 2D,3D,3D/3D/M/1.

We now describe some of the common forms of object recognition appearing in the literature.

### 4.1.1.1 *Image with Image*

Image with image fusion (2D,.../2D/*/*) refers to the combination of image data from several cameras, possibly at different wavelengths, observing the environment from different vantage points. The goal may be to construct a composite image by synthesizing the image data, or to reconstruct partial three-dimensional depth maps. Chipman et al. [36] describe the use of wavelets to fuse images from different wavelengths. Nandhakumar and Aggarwal [145] describe a method for object recognition using physics-based modeling of the image generation mechanisms to integrate image-derived information with model-derived information and the physics-based simulation of multisensory imagery.

### 4.1.1.2   *Vision and Range*

Vision and range fusion (2D,3D/3D/*/*) deals with the combination of two-dimensional image data with three-dimensional range data to reconstruct the workspace environment. Hel-Or and Werman [79] describe a method for fusing and integrating different 2D and 3D measurements for pose estimation. The 2D measured data is viewed as 3D data with infinite uncertainty in particular directions. The method is implemented using Kalman filtering (Section 3.3.1.1 on page 33). Umeda et al. [198] describe fusion of range image and intensity image for three-dimensional shape recognition. They fuse planar and cylindrical features extracted from a range image, with edge features from an intensity image, to achieve three-dimensional shape recognition.

### 4.1.1.3   *Vision and Touch*

Vision and touch fusion (2D,3D/3D/*/*) in a robot workspace deals with the combination of two-dimensional image data with three-dimensional touch data for object recognition and object manipulation. Manipulation of an object by only using visual information often falls into difficulties because of occlusions caused by the grasped object, by surrounding objects, and by the manipulator itself. Use of additional sensors, such as touch, force, or grasp can provide complementary information to aid in manipulation. Ishikawa et al. [88] describe estimation of contact position between a grasped object and the environment based on sensor fusion of vision and force data. Namiki and Ishikawa [144] describes optimal grasping using visual and tactile feedback. Kinoshita et al. [111] describe another approach fusing visual and tactile data. In Chapter 11 we describe the fusion of vision, touch, and grasp data for three-dimensional object recognition, to aid in robot manipulation.

## 4.2  Interpretation Techniques in Object Recognition

In Chapter 3 we discussed various techniques for multisensor data fusion. Many of these techniques are applicable, and have been used in object recognition. In the rest of this chapter, we discuss additional techniques specific to multisensor object recognition.

### 4.2.1  *Object Localization Approaches*

#### 4.2.1.1  *Clustering and Hough Transform*

Clustering and the related Hough transform approaches are based on the idea that pairings of subsets of model and data features can be used to compute a pose transform or a range of pose transforms, consistent with the pairing. Each pose transform can be plotted as a point in the parameter space of all possible pose transforms. The correct pairings will result in the same or similar pose transform, and thus a single large cluster will be formed in the parameter space. This cluster essentially defines the correct pose transform consistent with all the data. The object identity is usually assumed to be known, though the correspondence is unknown and is seldom computed explicitly in these methods. There is no explicit way of ensuring that only one data feature is paired to a model feature when such is the case.

Clustering and Hough transform differ in the way they implement this idea. Clustering typically uses traditional techniques, such as $k$-means, from the pattern recognition [48, 62] literature to find cluster centers in the parameter space. The difficulty with these methods is the definition of a suitable distance metric in the parameter space, since the position and orientation parameters are not directly comparable. Stockman and Chen [187], Stockman and Estava [188] describe a pose clustering method for three-dimensional objects restricted to lie on a planar surface, such as a table top, so that there are only three degrees of freedom (2D/3D/1/1). They use traditional k-means clustering on the three-dimensional parameter space.

Mundy and Heller [139] describe an object recognition system (2D/3D/1/1) that divides the six-dimensional transform space into four smaller subspaces, the output of one used as the input of the next, to make the clustering easier. First the orientation parameters are clustered, followed by position parameters; therefore they do not need to define distance functions in the entire transform space. A pose transform is generated by pairing a model vertex-pair with an image vertex-pair. To reduce the number of pose transforms generated they use a heuristic method, based on a rotation error measure, and restrict attention to a subset of model vertex-pair features. A large number of candidate clusters are formed, that must be pruned based on the support provided by the image data. The pose of the candidate clusters is used to project the model back into the image data for verification. The projection is used to determine an edge coverage value, which is the fraction of the predicted model boundary perimeter actually covered by the extracted image segments, and is used to select the best pose from the candidate clusters. They compute a distance transform on the image to make edge coverage computation an efficient process.

The Hough transform is implemented by subdividing or quantizing the parameter space into *bins*. A counter associated with each bin that keeps track of the number of pose transforms falling into it consistent with some pairing of the model and data. The bins with the largest counts define the clusters. A difficulty with this method is the large memory requirements when the parameter space is multidimensional.

Linnainmaa et al. [122] use a Hough transform to determine the six degree of freedom pose transform for three-dimensional objects (2D/3D/1/1). The entire Hough space is six-dimensional; its three-dimensional projection representing the translational parameters is used for practical reasons. A pose transform is generated by matching a triple of points on the object model to a triple in the image. The number of matches, or points in the Hough space, is reduced by using geometric constraints based on visibility and the uniqueness of imaging (i.e. a unique one-to-one correspondence exists). These heuristics are able to reject some spurious points and clusters in the Hough space. The pose represented by each cluster is found by taking a weighted average of all the poses in the cluster. The candidate pose transforms are used to project the object model back into the scene for verification. The pose which gives the least median distance between the projected model and the actual image is chosen to be the best one. They also discuss a least squares method to further improve the pose estimate, based on pose clustering results.

These techniques are based on the assumption that the sensor data is clean, and it works well when this is the case [122]. However the performance deteriorates with noisy data or spurious data. When the sensor data is relatively clean, there is one big dominant cluster. However as the noise in the data increases the clusters get diluted and the histogram is more spread out over the parameter space. It is hard to pick up one cluster as the dominant one. With spurious data or outliers present, the number of incorrect points in the parameter space increases and may give rise to false peaks or clusters. Grimson and Huttenlocher [70] have conducted an analysis of the performance of these methods, that confirms the limitations of this approach for complex scenes where the data is very noisy and the majority of it is outliers. However, this approach may still be useful as a preprocessing step (as a filter) to restrict attention to a smaller subset of the parameter space, which can then be evaluated by some other means [139].

Stockman and Estava [188] also discuss "cluster-pose-stereo" referring to clustering of pose parameters inferred from several camera images in a single parameter space, without having to perform any correspondence between the images (2D,2D/3D/1/1). This idea can be generalized to do multisensor fusion—cluster the parameters derived from each sensor data set into a single parameter space. Data across multiple sensors may be paired (i.e. one feature taken from sensor 1, another from sensor 2, etc. ) to arrive at more points in the parameter space. However this method suffers from the limitations, discussed above, in noisy and cluttered environments.

### 4.2.1.2 *Transform Space Geometry*

Another idea closely related to the clustering approach is to exploit the geometry of the transform parameter space. A *bounded error model* is used for the data—viz. a data feature can be anywhere in a certain bounded region around a model feature. Pairing a model feature and a data feature *constrains* the possible transforms in the parameter space and defines a region where the bounded error model is satisfied. This region is called the called a *match-region* [29]. Each pairing of model and data features reduces the space of feasible transforms (similar to the tree approach). The intersection of all the match regions is the set of transform parameters consistent with all the observed data. In practice, the region of the parameter space where the maximum number of match-regions overlap or intersect, is the solution region. Outliers can be rejected with this approach, and it finds a globally consistent solution.

Cass [29] describes a parallel implementation for matching contour features derived from the image data with model contour features (2D/2D/1/1). Each match-region is sampled and a global criterion evaluated that counts the number of match-regions overlapping at that sampled transform point. The point in the transform space with the highest evaluation is taken to be the best pose with this method. However, regions of maximal overlap can be very small and missed by the sampling.

Baird [9] shows that if the bounded error model is described by linear constraints, then for certain transformations they define linear constraints on the set of feasible transformations. Thus, polyhedral error bounds on the data give rise to polyhedral match-regions. Baird develops a linear programming solution to find the region of maximal overlap of the match regions. Breuel [24] uses this observation about linear constraints to develop an algorithm that recursively subdivides a given rectangular box in the transformation space, to identify portions where a large number of match-regions intersect (2D/2D/1/1).

### 4.2.1.3  *Hypothesize and Test*

Huttenlocher and Ullman [85] use *alignment* as a basis for recognizing three-dimensional solid objects in single two-dimensional images (2D/3D/1/1). Each possible alignment of the triples of model and image points yields a pose transform, which is computed using an affine approximation of the perspective projection. Pairs of edge features (each edge has a location and an orientation attribute) in the model/image are intersected to generate triples of point features, which are then aligned to hypothesize a pose. The hypothesized pose is verified by comparing entire edge contours of the transformed object model with the image edges. Image features are sorted so that transformed model features can be looked up in the image in logarithmic time.

Chen and Kak [34] describe their 3D-POLY system for pose estimation from range data in terms of a tree based approach, and use it for 3D/3D/M/M object recognition problems. They select a fixed size subset of the data features as the Hypothesis Generation Feature set (HGF). A pose can be hypothesized by pairing data features from the HGF with model features. The model features are organized or grouped into Local Feature Sets (LFS). Each each LFS a minimal grouping of features capable of yielding a unique pose transform. A model LFS is paired with the

HGF (the data LFS) to hypothesize a pose transform using a least squares method in constant time. The model features organized into a feature sphere—a spherical hash table obtained by tessellating a three-dimensional unit sphere. The hypothesized pose transform is used to project the data features back into the model feature sphere. A match is accepted when the percentage of matched data features exceeds a threshold, and the algorithm stops at the first match. The method relies heavily on the low-level processing producing good data features. The complexity of the algorithm for the 3D/3D/1/1 case is claimed to be quadratic (i.e. $O(n^2)$) in the number of data features $n$—there are roughly $O(n)$ model LFS, one scene LFS (the HGF), and it takes $O(n)$ time to verify the remaining data features using the feature sphere, once a pose is hypothesized. They do not address how the HGF (the data LFS) is selected; the number of possible permutations between a model LFS and a data LFS contributes a constant factor in the complexity calculation and so drops out. The savings come from the organization of the model features into LFS's and the use of a feature sphere for verification.

### 4.2.2  *Object Matching Approaches*

#### 4.2.2.1  *Relaxation Labeling*

Relaxation labeling [83] is a common iterative technique applied to matching problems. A match quality is associated with each node representing a match between a model and a data feature. In *discrete* relaxation the quality of match is binary, while in *continuous* relaxation it is a real valued number between 0 and 1. The overall match is a global property and is obtained by *iteratively* performing local updates on the match quality, based on local compatibility measures between the node and its immediate neighbors. This is repeated for all the nodes per cycle. At the end of each cycle a new solution emerges, and hopefully converges to a stable state after a number of iterations. Convergence proofs are known only for some special cases, and in general this process can get stuck in local minima or may not converge.

#### 4.2.2.2  *Graph Matching*

One way of establishing the correspondence between model and data features sets is to form a graph of pairwise consistent model and data features, and then search for the largest clique. This is the approach taken by Bolles and Cain [20] in their Local Feature Focus (LFF) system. The maximal clique is taken to be the best correspondence, although there is no real reason to be so. Furthermore, the clique finding problem is NP-complete.

Kim and Kak [110] use bipartite graph matching to establish the correspondence between model and scene features. They develop the method for the 3D/3D/M/1 problem, and use it for the 3D/3D/M/M problem when the scene can be segmented into different objects. The model and data features are represented as a relational graph—each graph node representing a feature, and the arcs representing representing the relation between features. The matching problem is to establish a one-to-one correspondence between the model and the data features. A bipartite graph is constructed with the model nodes on the left and the data nodes on the right and arcs linking the consistent nodes. Initially bipartite matching is used to find one feasible matching, which is used to construct an assignment table specifying the consistent nodes. Discrete relaxation is then used on the assignment table to enforce relational constraints, by making sure that the neighbors of consistent nodes are also consistent and so on. This step is also implemented using bipartite graph matching. Once a match is found, the pose transform is computed using a least squares method. The computed pose is used to transform the model and verify the match. The work does not address handling of spurious features in the data, although it is applied to finding multiple object models in the scene. It depends heavily on the ability of the low-level processing to generate a good segmentation of the scene.

### 4.2.2.3  *Hopfield Neural Nets*

Lin et al. [121] use a Hopfield neural net [82] to find the correspondence between object features and data features (2D/3D/1/1). The polyhedral object model is represented using *aspect graphs* [65], with each node representing a *characteristic view* [33]. Each characteristic view is matched to the image. First the surface features are matched, and for the matched surfaces the vertex features are matched next. The image features are represented as the rows of the Hopfield net, while the object features are represented as the columns. Matching is regarded as a constraint satisfaction problem, and an energy function is is set up in terms of the compatibility of the n-odes. The network is allowed to converge to a stable state, the active nodes (neurons) indicating the match.

Nasrabadi and Li [147] take a similar approach for two-dimensional object models, with multiple objects competing simultaneously for a match in the image (2D/2D/M/M). The matching results are clustered to compute the pose, which is then used to eliminate the wrong matches. Nasrabadi and Choo [146] use a similar technique to find the correspondence between *interest points* in the left and right images for stereo vision.

Parvin and Medioni [153] use a Hopfield net to solve the correspondence problem for three-dimensional range images (3D/3D/1/1), by setting it up as a constraint satisfaction problem, and using a Hopfield neural net to solve it. Note that, the biggest drawback of these approaches is that the Hopfield neural net can get stuck in local minima. There is a resemblance to the relaxation labeling technique.

#### 4.2.2.4  *Invariants*

The use of invariants is popular in the computer vision literature. The idea is to compute some quantities from the data that are invariant with respect to the sensing model. Specifically invariants that are independent of the pose transformation are computed from the data and the model. These may be regarded as shape descriptors immune to a class of transformations, such as translations and rotations. The invariants can be used to classify the data into various object classes, or compute the correspondence between the object model and the data.

Subrahmonia et al. [192] describe an approach in which the data is segmented and surface patches described by implicit polynomials are fitted to each segment. Invariants are computed from the fitted surface patches described by algebraic polynomials. These invariants from the data are compared to the model invariants. Essentially this becomes a classification problem in the space of invariants—pick the model invariant nearest to the data invariant; and thus identify the object.

Allen and Roberts [6] recover superquadric object models from the data and then match their parameters to those of the models in the library.

The main weakness of an invariant based approach, is the sensitivity of the invariants to noise and outliers in the data. Developing invariants robust to noise and outliers is an open research problem.

### 4.2.2.5  *Symbolic and Geometric Reasoning Methods*

Symbolic reasoning has been applied to match objects models with data. The raw sensor data is processed to extract features, each feature described by a number of symbolic and numerical attributes as well as relational attributes. An object model is represented similarly in terms of features with associated attributes. Symbolic and geometric reasoning is applied to match the model with the data.

Walker et al. [205] take a knowledge based geometric reasoning approach in their 3D-FORM system. They use symbolic and geometric reasoning along with three-dimensional objects models, to interpret two-dimensional images. The ACRONYM system [25] developed by Brooks, uses constraint manipulation and symbolic reasoning to establish a correspondence between the data and the model features. Vayda and Kak [203] use geometric reasoning to perform generic object recognition from range data in their INGEN system.

### 4.2.2.6  *Tree Based Sequential Hypothesize and Test*

Grimson and Lozano-Pérez [71] presented a scheme (3D/3D/1/1) for recovering pose by building an *interpretation tree*. Model features are represented as nodes in the tree, with each arc emanating from a node representing a data feature (or rather a pairing of the model feature with a data feature). A special "NIL" arc indicates no data feature corresponding to that model feature. The resulting tree is called an interpretation tree (IT) [71], and a path in the tree denotes a matching between the model and the data features. Problem constraints are exploited to reject the impossible parings of features, until a finite set of feasible poses remain. If there is a relation $D_i \ R \ D_j$ in the data, then the relation $M_i \ R \ M_j$ should also hold in the model, when the model feature $M_i$ is matched to the data feature $D_i$ and $M_j$ to $D_j$. These local constraints are used to prune the IT until a set of valid matchings (paths) remain. The tree can be searched in a depth-first order. A path is not explored any further if a constraint is violated. Multiple paths may exist for symmetric objects. A pose can be computed from a subset of the matched features, and used to verify the the rest of the correspondences in the match. Poses are computed from the matches that survive all these tests, and clustered to find the object pose. In general, this approach would be combinatorially expensive. Siegel [179] used a similar scheme for finding the pose of two-dimensional objects grasped by a robot hand.

Faugeras and Hebert [56] use a similar tree based method for finding the best correspondence and pose transform (3D/3D/1/1). For every partial path in the tree (specifying a partial matching) a least squares pose transform estimate is computed. This pose transform is then used to transform the next unmatched model feature, and only consider as possible candidates those unmatched data features that are sufficiently close to it. This reduces the breadth of the search tree. Faugeras and Hebert refer to this scheme as the *recognizing while localizing paradigm*. They present exact and iterative solutions to the least-square pose estimation problem—the iterative method bears some semblance to extended Kalman filtering. The initial pose hypothesis is formed by aligning the first two pairings of the data and the model features of a path in the tree.

Stockman and Chen [187] describe a more general version of the hypothesize and test paradigm, and compare it with the pose clustering paradigm.

The RANdom SAmple Consensus, or the RANSAC algorithm, proposed by Fischler and Bolles [58] is also a variant of the sequential hypothesize and test approach (2D/3D/1/1). In the RANSAC paradigm, model parameters estimated from a subset of the data are used to project the model back to the data, and reject the datum (used in estimation) that exceed an error threshold as outliers. The resulting data set, called the consensus set, is used to compute a new model parameter estimate. For the pose estimation problem, three randomly selected model and data features are matched and estimated parameters used to evaluate their consistency. Consistent parings are added to the consensus set of matched model and data features. When the consensus set exceeds a size threshold, a least squares method is used to compute the final transform parameters.

Note that all these methods first establish a correspondence between the model and the data features, which is then used to compute the pose transform parameters.

# Chapter 5

# Minimal Representation

*All theories should be as simple as possible but no simpler!*
*—Albert Einstein*

## 5.1 Introduction

The statistical *minimal representation size* criterion, also known as the *minimum description length* criterion in the literature, has been proposed as a general criterion for model inference by Rissanen [165] and by Segen and Sanderson [177]. It is an expression of the ideas on algorithmic information theory pioneered by Solomonoff [183], Kolmogorov [113], and Chaitin [30]. It is related to the information measure used by Wallace and Boulton [206] and has been applied to problems in clustering [92, 108, 177, 206], image processing [28, 57, 117, 152, 215, 218], signal processing [207], attributed image matching [158, 173], shape matching [97], density estimation [13], learning [213], neural networks [16, 210], genetic programming [114], target tracking [86], and multisensor fusion [95, 98–104].

The minimal representation approach is based on the principle of building the shortest length program which reconstructs the observed data. The length of such a program or *representation size* includes terms for a stored model, a reconstruction algorithm, and an encoded representation of the residuals. According to the minimal representation principle for model inference, the best model of the data is the one with the smallest representation size. This approach provides an objective criterion to trade-off between model size, algorithm complexity, and observation errors.

This chapter provides a comprehensive introduction to the principles

57

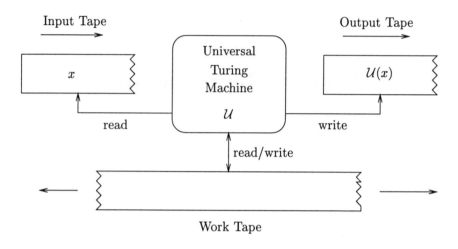

Fig. 5.1   Deterministic Universal Turing Machine $\mathcal{U}$. It reads a program $x$ from the input tape and produces a string $\mathcal{U}(x)$ on the output tape.

of minimal representation. We will begin by describing the notion of Kolmogorov complexity and show how it leads to the development of a practical statistic, summarized as the minimal representation principle. Next, we discuss the work on the statistical minimal representation criterion when stochastic models of the data are available. Most theoretical work and practical applications in the literature have focused on this version of the minimal representation principle. Finally, we discuss optimal and suboptimal encoding schemes for various commonly occurring entities, useful in developing a minimal representation criterion for various problem domains.

Other good sources of literature on minimal representation principles are [1, 12, 142, 165, 169, 177, 185].

## 5.2   Representation Size

The notion of *Kolmogorov complexity* [113], discussed at length in Cover and Thomas [40], is based on the idea of writing a program to generate observed data. Consider the deterministic Universal Turing Machine (UTM) [120], shown in Figure 5.1. It has a unidirectional read-only input tape, a unidirectional write-only output tape, and a bidirectional read/write work tape. The Turing machine, $\mathcal{U}$, reads a program $x$ from the input tape and produces a string $\mathcal{U}(x)$ on the output tape.

The observed data $\mathcal{D}$ is viewed as a bit string to be produced on the output tape of the UTM. A program that produces $\mathcal{D}$ on the output tape is called a *representation* of $\mathcal{D}$. Specifically, a program $x$ is a representation of $\mathcal{D}$ if, when executed on the UTM, it produces the string $\mathcal{D}$ as the prefix of the output $\mathcal{U}(x)$, i.e. $\mathcal{D} = \text{Prefix}[\mathcal{U}(x)]$. Thus, the representation size of some observed data is the *length* of a program in bits, that when executed on the UTM, would reproduce the observed data on the output tape (Figure 5.1).

The representation $x$ of data $\mathcal{D}$ is viewed as an encoding that can achieve data compression or compaction. Viewed this way, representation size is the number of bits needed to *decode* observed data. Restricting attention to finite length programs on which the UTM halts leads to the notion of Kolmogorov complexity of the string $\mathcal{D}$. It is easy to see that the programs that halt are *self-delimiting* (in the sense that no halting program can be a prefix of another halting program) and therefore satisfy the Kraft inequality [17, 40] for prefix codes

$$\sum_{x:\,\mathcal{U}(x)\text{ halts}} 2^{-|x|} \leq 1 \tag{5.1}$$

where $|x|$ is the length (in bits) of program $x$. The Kolmogorov complexity $\mathcal{L}[\mathcal{D}]$ of a string $\mathcal{D}$ is defined as the length of the most compact (the smallest size) representation of $\mathcal{D}$ in the set of all halting programs.

The entire subject of *algorithmic information theory* [31, 32] has developed around the notion of Kolmogorov complexity, and is indeed a general framework [30] encompassing classical information theory [17, 40]. However, there is a practical difficulty associated with using Kolmogorov complexity as a statistic for model inference—it has been shown that, in general, the Kolmogorov complexity is *uncomputable*. Given a representation $x$, deciding whether it is the most compact representation or not is an unsolvable problem.

A practical resolution of this predicament is to restrict attention to a class of input programs, $X$, that form a subset of the set of all halting programs (Figure 5.2). The length of the most compact representation of $\mathcal{D}$ with respect to a class of programs $X$, denoted as $\mathcal{L}[\mathcal{D}\,|\,X]$, defines a computable measure of complexity with respect to the class $X$ (provided $X$ is "appropriately" chosen). We might call $\mathcal{L}[\mathcal{D}\,|\,X]$ the *relative Kolmogorov complexity*. This approach leads to the minimal representation method [12,

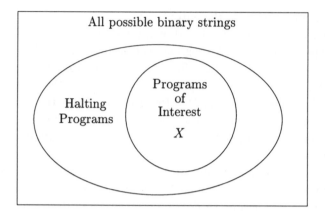

Fig. 5.2   The minimal representation approach. Restricting attention to a subset $X$ of the halting programs can lead to a computable measure of relative complexity with respect to the class $X$.

166, 177], where a library $\mathbf{Q}$ of the models is available and the data is encoded in terms of the models in the library. Implicit is the assumption that there exists a computable *isomorphism* between the encoding of data using the models in the library and some class $X(\mathbf{Q})$ of halting programs.

In a model-based coding scheme, observed data is thought to be arising from *one* of the several available models in a model library $\mathbf{Q}$. Given the model library, the data $\mathcal{D}$ is encoded in terms of a model $\mathbf{q} \in \mathbf{Q}$ and the residuals of the data $\mathcal{D}$ with respect to the selected model $\mathbf{q}$. Using such a "two-part" encoding scheme,

$$\mathcal{L}[\mathcal{D} \mid \mathbf{Q}] \triangleq \min_{\mathbf{q} \in \mathbf{Q}} \mathcal{L}[\mathbf{q}, \mathcal{D} \mid \mathbf{Q}] = \min_{\mathbf{q} \in \mathbf{Q}} \mathcal{L}[\mathbf{q} \mid \mathbf{Q}] + \mathcal{L}[\mathcal{D} \mid \mathbf{q}, \mathbf{Q}]$$

where $\mathcal{L}[\mathcal{D} \mid \mathbf{Q}]$ denotes the *representation size* of the data $\mathcal{D}$ given the model library $\mathbf{Q}$. $\mathcal{L}[\mathbf{q} \mid \mathbf{Q}]$ is the number of bits needed to represent the model itself, and $\mathcal{L}[\mathcal{D} \mid \mathbf{q}, \mathbf{Q}]$ is the number of bits needed to encode the data given the model $\mathbf{q}$. As a shortcut, we often write the last two terms as $\mathcal{L}[\mathbf{q}]$, and $\mathcal{L}[\mathcal{D} \mid \mathbf{q}]$ respectively, with the understanding that there in an underlying library $\mathbf{Q}$ relative to which these terms are computed.

Different encoding algorithms may be used to encode the data with respect to a given model. The choice of an encoding algorithm will lead to a specific codelength of the data residuals. In order to regenerate observed

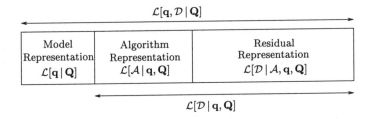

Fig. 5.3   Representation size $\mathcal{L}[\mathbf{q}, \mathcal{D} \mid \mathbf{Q}]$ of observed data $\mathcal{D}$ encoded in terms of a model q from a library $\mathbf{Q}$, using an encoding algorithm $\mathcal{A}$.

data the encoding algorithm $\mathcal{A}$ must also be specified. Thus,

$$\mathcal{L}[\mathcal{D} \mid \mathbf{q}, \mathbf{Q}] = \mathcal{L}[\mathcal{A} \mid \mathbf{q}, \mathbf{Q}] + \mathcal{L}[\mathcal{D} \mid \mathcal{A}, \mathbf{q}, \mathbf{Q}]$$

where $\mathcal{L}[\mathcal{A} \mid \mathbf{q}, \mathbf{Q}]$ is the number of bits needed to describe the encoding algorithm, and $\mathcal{L}[\mathcal{D} \mid \mathcal{A}, \mathbf{q}, \mathbf{Q}]$ is the number of bits needed to encode the data residuals given a model and a specific coding algorithm. Combining the previous two expressions, the representation size of data $\mathcal{D}$, given a model library $\mathbf{Q}$ is expressed by

$$\mathcal{L}[\mathcal{D} \mid \mathbf{Q}] \;=\; \min_{\mathbf{q} \in \mathbf{Q}} \mathcal{L}[\mathbf{q} \mid \mathbf{Q}] \;+\; \mathcal{L}[\mathcal{A} \mid \mathbf{q}, \mathbf{Q}] \;+\; \mathcal{L}[\mathcal{D} \mid \mathcal{A}, \mathbf{q}, \mathbf{Q}]$$
$$\text{Rep. Size} \;=\; \text{Model Size} \;+\; \text{Algorithm Size} \;+\; \text{Residual Size}$$
$$(5.2)$$

and is illustrated in Figure 5.3.

## 5.3   The Minimal Representation Principle

The previous discussion leads us to the minimal representation principle.

**Proposition 5.1**   *Given a library $\mathbf{Q}$ of models that can explain observed data, the best model, $\mathbf{q}_{opt}$, is the one with the smallest representation size:*

$$\mathbf{q}_{opt} = \arg\min_{\mathbf{q} \in \mathbf{Q}} \mathcal{L}[\mathbf{q} \mid \mathbf{Q}] + \mathcal{L}[\mathcal{A} \mid \mathbf{q}, \mathbf{Q}] + \mathcal{L}[\mathcal{D} \mid \mathcal{A}, \mathbf{q}, \mathbf{Q}] \qquad (5.3)$$

*where $\mathcal{L}[\mathbf{q} \mid \mathbf{Q}]$ is the model representation size, $\mathcal{L}[\mathcal{A} \mid \mathbf{q}, \mathbf{Q}]$ is the algorithm representation size, and $\mathcal{L}[\mathcal{D} \mid \mathcal{A}, \mathbf{q}, \mathbf{Q}]$ is the residual representation size given the model and the encoding algorithm.*

A model inference rule based on this principle is called a *minimal representation criterion*. . It chooses the minimum complexity solution relative to the model library **Q**, and is an objective criterion to trade-off between model size, algorithm complexity, and observation errors. Note that representation size is a *sufficient statistic* [184] for model inference, since the minimal representation can be decoded by the UTM to regenerate observed data.

The minimization in Equation (5.3) is over the models in the library. The models may have various structures, and the model structures may be parameterized, each with a different number of parameters. The only requirement is that they form a countable set.

The choice of the algorithm $\mathcal{A}$ that encodes the data in terms of the model is left open. Thus, one would like to use an *optimal encoding scheme* which gives the shortest codes (often on an average) for the data residuals. This is an open research question, not answered by the minimal representation principle. It leaves room for subjective and domain specific interpretation; the "optimal" answer would depend on the nature of the problem and as well as the choice of models constituting the library. Choosing a specific coding scheme would lead to a specific minimal representation *criterion*. Developing optimal coding schemes for model-based multisensor fusion constitutes one of the major contributions of this work.

Several arguments can be used to arrive at the minimal representation principle. Interpreting representation size as generalized algorithmic entropy, this principle finds the minimum entropy solution i.e. one which achieves the best data compaction. Sorkin [185] arrives at the same conclusion, basing his arguments on statistical mechanics and using the notion of thermodynamic entropy to define priors for "phenomenological" theories, or models, as we call them here.

The minimal representation principle can also be interpreted as a quantitative formalization of the *Occam's Razor* [91, 185]

"Plurality is not to be assumed without necessity"

an assertion made by the 14-th century English philosopher, William of Ockham. It says that choosing simpler or smaller models is better than choosing larger ones, unless there is a compelling reason to use a larger model. In the context of representation size, a simple model (or no model) would require describing the data in great detail and the data residual would be large. A large or complex model would require little further elaboration and the data residual would be small. The minimal representation principle gives a criterion that balances the trade-off between model size and , residual size, choosing a larger model only when necessary.

Segen and Sanderson [177] present the minimal representation principle using a Bayesian argument which proceeds as follows. The idea is to assign lower probabilities to more complex models. In the UTM of Figure 5.1, assume the input string comes from a completely random or maximum entropy source. Since there is no prior assumption about the input program $x$, it can be thought of as being generated by independent random tosses of a fair coin; and can be viewed as an extension of the *principle of insufficient reason* [183]. Therefore, the *a priori* probability of a program $x$ is

$$P(x) = 2^{-|x|},$$

where $x$ is a halting program and therefore satisfies the Kraft inequality (Equation 5.1). Consider the class of representations of the observed data $\mathcal{D}$

$$\mathcal{S}(\mathcal{D}) = \{x \mid \mathcal{D} = \text{Prefix}[\mathcal{U}(x)]\}.$$

Given a program $x$, the conditional probability that $\mathcal{D}$ is produced on the output tape is

$$P(\mathcal{D} \mid x) = \begin{cases} 1 & \text{if } x \in \mathcal{S}(\mathcal{D}) \\ 0 & \text{otherwise} \end{cases}$$

Now, consider the subset of representations that belong to a specified class $X$ of programs, i.e. consider the set of representations $\mathcal{S}(\mathcal{D}) \cap X$. The conditional probability of $x$ given $x \in X$ is

$$P(x \mid X) = \frac{P(x \cap X)}{P(X)} = \begin{cases} \frac{2^{-|x|}}{P(X)} & \text{if } x \in X \\ 0 & \text{otherwise} \end{cases}$$

Using Bayes rule [151], the *a posteriori* probability of program $x$, given that $\mathcal{D}$ appears on the output tape and $x \in X$ is

$$P(x \mid \mathcal{D}, X) = \frac{P(x, \mathcal{D}, X)}{P(\mathcal{D}, X)} = \frac{P(\mathcal{D} \mid x, X) \, P(x, X)}{P(\mathcal{D} \mid X) \, P(X)} = \frac{P(\mathcal{D} \mid x, X) \, P(x \mid X)}{P(\mathcal{D} \mid X)}$$

$$\propto \begin{cases} 2^{-|x|} & \text{if } x \in \mathcal{S}(\mathcal{D}) \cap X \\ 0 & \text{otherwise} \end{cases}$$

$$(5.4)$$

Thus, given the data string $\mathcal{D}$ and a class of programs $X$, the *most likely* input program $x_{\text{opt}}$ is the minimal length representation of $\mathcal{D}$ in $X$, i.e.

$$|x_{\text{opt}}| = \min_{x \in \mathcal{S}(\mathcal{D}) \cap X} |x|.$$

This is the minimal representation principle. In Equation (5.3), the model library is understood to induce a class of programs $X(\mathbf{Q})$. Therefore, Equation (5.3) is an alternative way of expressing the above conclusion.

## 5.4   Statistical Model Inference

Barron et al. [12], Rissanen [165, 169] restricted attention to stochastic models described by probability distributions. Segen and Sanderson [177] also concentrate on the case of statistical models and develop the method for stochastic model inference when the number of parameters is unknown. We discuss this special case next. However, note that the minimal representation principle can be used with other kinds of models; we take this more general outlook in developing a technique for multisensor fusion.

For the case of stochastic models considered by Rissanen and by Segen and Sanderson, the library $\mathbf{Q}$ is a collection of families of *discrete* probability distributions, each family with a different structure and number of parameters. A model $\mathbf{q} \in \mathbf{Q}$ specifies the structure of the probability distribution along with its discrete parameters, and describes the data distribution $P_{\mathbf{q}}(\mathcal{D})$. The model inference problem is to estimate the model structure (the number of parameters) and the parameters values of the probability distribution.

Segen and Sanderson [177] constructed a computable isomorphism $v$ from the class of models $\mathbf{Q}$ to a class $X$ of programs,

$$v : \{(\mathbf{q}, \mathcal{D}) : P_{\mathbf{q}}(\mathcal{D}) > 0, \ \mathbf{q} \in \mathbf{Q}\} \to X(\mathbf{Q})$$

such that if $v(\mathbf{q}, \mathcal{D}) = x$, then

$$\mathcal{L}[\mathbf{q}, \mathcal{D} \mid \mathbf{Q}] = L + \mathcal{L}[\mathbf{q}] + \lceil -\log_2 P_{\mathbf{q}}(\mathcal{D}) \rceil = |x| \qquad (5.5)$$

and

$$\mathcal{L}[\mathcal{D} \mid \mathbf{Q}] = \mathcal{L}[\mathcal{D} \mid X]$$

where $\mathcal{L}[\mathcal{A} \,|\, \mathbf{q}, \mathbf{Q}] = L$, the (fixed) length of *Shannon coding* algorithm, and $\mathcal{L}[\mathbf{q} \,|\, \mathbf{Q}] = \mathcal{L}[\mathbf{q}]$, the number of bits needed to encode the model. Although they do not consider the "optimality" of this coding algorithm, it is easy to see that Shannon codes achieve the shortest code lengths on an average (since the expected value is the entropy when the true distribution is used). Rissanen [165], using a slightly different reasoning, arrived at an expression similar to Equation (5.5), for the representation size of stochastic data. Using *Gibbs theorem* he argued that this expression gives the shortest data encoding when the probability distributions are known. He also presented a *combinatorial entropy* argument to justify this expression. Thus, Equation (5.5) is the representation size for an optimal encoding of data using stochastic models.

The minimal representation criterion for *statistical model inference* is obtained by using Equation (5.5) for representation size and dropping the ceiling

$$\boxed{\mathbf{q}_{\text{opt}} = \min_{\mathbf{q} \in \mathbf{Q}} \mathcal{L}[\mathbf{q}] - \log_2 P_{\mathbf{q}}(\mathcal{D})} \qquad (5.6)$$

This criterion can be seen as a tradeoff between *simplicity* and *likelihood* of models describing the observed data: the simplest model (i.e. smallest $\mathcal{L}[\mathbf{q}]$) that maximizes the likelihood (i.e. $P_{\mathbf{q}}(\mathcal{D})$) of the observed data is chosen. It is noteworthy that Equation (5.6) admits a Bayesian interpretation, if the prior probability of $\mathbf{q}$ is taken to be $P(\mathbf{q}) = 2^{-|\mathcal{L}[\mathbf{q}]|}$. Then, Equation (5.6) is seen to be equivalent to maximizing the posterior probability of $P(\mathbf{q} \,|\, \mathcal{D})$. However, our development of this criterion is based on the more general minimal representation principle, and not restricted to this narrower choice of a prior distribution.

Given $N$ observed data features $\mathbf{z}_1, \dots, \mathbf{z}_N$, we write $\mathcal{D} = \mathbf{z}_1, \dots, \mathbf{z}_N$ where $\mathbf{z}_i$ may be a $n$-dimensional feature vector $\mathbf{z}_i = [z_{i1}, \dots, z_{in}]^{\mathrm{T}}$. The probability distribution of the observed data is the joint distribution of all the observed features $P_{\mathbf{q}}(\mathcal{D}) = P_{\mathbf{q}}(\mathbf{z}_{i1}, \dots, \mathbf{z}_{in})$, which simplifies to $P_{\mathbf{q}}(\mathcal{D}) = \prod_{i=1}^{N} P_{\mathbf{q}}(\mathbf{z}_i)$ when the data features are independent and identically distributed (i.i.d.). Therefore, the statistical minimal representation criterion of Equation (5.6) simplifies to

$$\boxed{\mathbf{q}_{\text{opt}} = \min_{\mathbf{q} \in \mathbf{Q}} \mathcal{L}[\mathbf{q}] - \sum_{i=1}^{N} \log_2 P_{\mathbf{q}}(\mathbf{z}_i)} \qquad (5.7)$$

for i.i.d. data features, and is the form most commonly used in practical applications.

Most work on minimal representation has focused on this statistical minimal representation criterion (Equations 5.6, 5.7). Rissanen [167] showed that it achieves the best data compression in the class of universal codes. Barron and Cover [13] showed that such a minimum complexity probability distribution is statistically accurate and the rate of convergence is comparable to other methods of parametric and non-parametric estimation. They also proved that it leads to *consistent* [184] model estimates which converge to the true distribution when it belongs to the model library. More theoretical results can be found in Barron et al. [12], Barron and Cover [13], Najmi et al. [142], Rissanen [169], Rissanen et al. [170], Segen and Sanderson [177].

We have used the statistical minimal representation criterion to determine the optimal number of clusters [92]. It has been successfully used to find consistent order estimates of AR and ARMA models [169, 177]. In our previous work [95, 97–104], we have structured the model-based pose estimation problem such that the pose transformation parameters are isolated elements of the statistical model, and are estimated by the minimal representation procedure.

## 5.5   Encoding Schemes

In this section we discuss optimal and sub-optimal encoding schemes for commonly occurring objects: sets, integers, real numbers, finite intervals, and probability densities. These are useful in developing representation size expressions for different problem domains, and also provide insight into model encoding, $\mathcal{L}[\mathbf{q}]$.

An encoding scheme is considered optimal, if it produces the shortest codes (at least on an average) for the data residuals $\mathcal{L}[\mathcal{D} \mid \mathbf{q}]$ given a model $\mathbf{q}$. While devising encoding schemes, one must ensure that the resulting code is unambiguously decodable. Self-punctuating prefix codes must satisfy the Kraft inequality (Equation 5.1). It may not always be straightforward to devise optimal self-punctuating codes.

### 5.5.1 *Finite Sets*

An element of a finite set $S$ containing $|S|$ elements is described by

$$\mathcal{L}[s \in S \,|\, S] = \lceil \log_2(|S| + 1) \rceil \approx \log_2(|S| + 1) \quad bits \qquad (5.8)$$

assuming that the UTM knows the set $S$.

### 5.5.2 *Integers*

Let us denote the representation size of a *non-negative* integer $n$ by $\mathcal{I}(n)$, i.e. define

$$\mathcal{I}(n) \overset{\Delta}{=} \mathcal{L}[n \,|\, n \geq 0].$$

We consider two ways of writing uniquely decodable *prefix* codes for non-negative integers and develop expressions for $\mathcal{I}(n)$.

In the first method, described by Segen and Sanderson [177], a non-negative number $n$ is represented in *ternary*, using 2 bits for each ternary digit: $(00) = 0$, $(01) = 1$, $(10) = 2$; the remaining code $(11)$ indicates the end of the number. Thus, for a non-negative integer $n$, we have

$$\mathcal{I}(n) = 2\lceil \log_3(n + 1) \rceil + 2 \quad bits \qquad (5.9)$$

as the representation size of integer $n$. We often drop the ceiling and use

$$\mathcal{I}(n) = (2 \log_3 2) \log_2(n + 1) + 2 \approx 2 + 1.26 \log_2(n + 1) \quad bits \qquad (5.10)$$

in practical applications.

In the second method, described by Rissanen [166, 169], the idea is to encode a non-negative integer $n$ by its binary representation using $\lceil \log_2(n + 1) \rceil$ bits. However the decoder must also be told the length of this representation, and specifying that requires $\lceil \log_2(\lceil \log_2(n + 1) \rceil + 1) \rceil$ bits. But now the length of this length must be specified, and so on. Dropping the ceilings, we require about $\log_2(n + 1) + \log_2 \log_2(n + 1) + \ldots$ bits to encode $n$ where the sum is over the positive iterates. Define the *iterated logarithm* of a number $n$ as

$$\log^* n \overset{\Delta}{=} \log n + \log \log n + \ldots$$

where only positive terms are included in the sum. The representation size of a non-negative integer is defined to be

$$\mathcal{I}(n) = \log_2 c + \log_2^*(n+1) \quad bits \tag{5.11}$$

where the constant $c$ is determined so as to satisfy the Kraft inequality exactly $\sum_{n=0}^{\infty} 2^{-\mathcal{I}(n)} = 1$, which yields $c = 2.865064$. This ensures we don't "waste" any codewords in the infinite binary tree describing all the prefix codes. Rissanen [166] showed that Equation (5.11) can be taken as the "just about" ideal code length for large positive integers, in the sense that we can do no better (in the mean) even if we had natural prior on the integers to design the optimal code. Furthermore, the first term of $\log_2^*(n+1)$ is the dominant one, and we may approximate the Equation (5.11) as

$$\mathcal{I}(n) \approx \log_2 c + \log_2(n+1) \approx 1.52 + \log_2(n+1) \quad bits \tag{5.12}$$

In this approximation, we are omitting the number of bits needed to specify the length of the binary *representation* of $n$.

In practice, we use a combination of the two encoding techniques: a non-negative integer is encoded using $\log_2(n+1)$ bits, and the length of this encoding is encoded using $1.26 \log_2(\log_2(n+1)+1)+2$ bits, resulting in a total of

$$\boxed{\mathcal{I}(n) \approx \log_2(n+1) + 1.26 \log_2(\log_2(n+1)+1) + 3.52 \quad bits.} \tag{5.13}$$

This gives an excellent approximation to the optimal codelength of $n$. The encoding of integers objects that can take on positive or negative values is done by encoding their magnitude and appending a *sign bit*. Thus,

$$\mathcal{L}[n] = 1 + \mathcal{I}(|n|) \tag{5.14}$$

where $n$ may be positive or negative.

### 5.5.3  *Probability Densities*

Stochastic models were introduced in Section 5.4. For *discrete* probability distributions $P_\mathbf{q}(\mathbf{z})$ on the data, the optimal coding scheme leads to a representation size of

$$\mathcal{L}[\mathbf{z} \,|\, \mathbf{q}, \mathcal{A}_{\text{opt}}] = -\log_2 P_\mathbf{q}(\mathbf{z}) \quad bits$$

where $\mathcal{A}_{opt}$ is the Shannon encoding algorithm, which produces the shortest *mean* code length for the data residuals. It is easy to see that the *average* value of this term is *entropy* in statistical information theory [40].

Now consider the case when we have a *continuous* distribution on the data, with a probability *density* $p_q(z)$. We discretize the continuous random variable into quantization intervals of width $\delta$, where $z$ and $\delta$ may be $n$-dimensional vector quantities. Thus,

$$\Pr\{z - \frac{\delta}{2} \leq \xi < z + \frac{\delta}{2} \,|\, q\} = \int_{z \pm \frac{\delta}{2}} p_q(\xi)\, d\xi_1 \ldots d\xi_n = p_q(\bar{z}) \prod_{k=1}^{n} \delta_k$$

where the second term follows from the *mean-value* theorem of integral calculus and $\bar{z}$ is a representative point in the quantization interval, $z - \frac{\delta}{2} \leq \bar{z} < z + \frac{\delta}{2}$.

Consider the discrete probability distribution defined on the representative points $\bar{z}_i$ over all the quantization intervals. We associate a probability mass $p_q(\bar{z}_i) \prod_{k=1}^{n} \delta_k$ with the point $\bar{z}_i$, and encode the resulting discrete probability distribution. The results of Section 5.4 are applicable to it, and the representation size of observed data residuals is approximated as

$$\mathcal{L}[z \,|\, q, \mathcal{A}_{opt}, \delta] = -\log_2 p_q(z) - \sum_{k=1}^{n} \log_2 \delta_k \quad bits \qquad (5.15)$$

and holds exactly in the limit as $\delta \to 0$.

In practice, we sometimes drop the term encoding the quantization interval $\delta$ in Equation (5.15), since it contributes a constant term to the minimal representation criterion. Thus, we sometimes use

$$\mathcal{L}[z \,|\, \mathcal{A}_{opt}, q, \delta] = -\log_2 p_q(z) \quad bits$$

which may produce negative values for representation size. This simply means the representation size surface is *shifted* by a constant amount to be below zero. One might call this the *differential* representation size, in analogy with the notion of differential entropy [40] for continuous random variables. In Equations (5.6) and (5.7), $P_q(z)$ can be replaced by $p_q(z)$, and the resulting criterion used for statistical model inference.

### 5.5.4  *Reals*

We encode the real numbers by discretizing them and converting into integers. Let us denote the representation size of a *non-negative* real number $u \in \Re, u \geq 0$, discretized to a quantization interval $\delta$ by

$$\mathcal{B}(u; \delta) \triangleq \mathcal{L}[u \mid u \in \Re^+, \delta].$$

Dividing $u$ by the quantization interval $\delta$ and truncating, we get the integer $\lfloor \frac{u}{\delta} \rfloor$. Encoding this integer gives the representation size of $u$ at a discretization $\delta$,

$$\mathcal{B}(u; \delta) = \mathcal{I}(\lfloor \frac{u}{\delta} \rfloor) \quad bits. \tag{5.16}$$

We can use either of the expressions of Equation (5.9) or (5.11) to compute $\mathcal{B}(u; \delta)$. Substituting in the expression for $\mathcal{I}(\cdot)$, we get the approximation

$$\mathcal{B}(u; \delta) = \mathcal{I}(\lfloor u + \delta \rfloor) - \log_2 \delta \quad bits \tag{5.17}$$

where the right hand side is always $\geq 0$. This expression is analogous to Equation (5.15) for continuous distributions. Also notice that Equation (5.17) follows from Equation (5.15), if $2^{-\mathcal{I}(\lfloor u \rfloor)}$ is regarded as a prior distribution of the random variable $u \in \Re^+$ quantized to an interval $\delta$.

For real numbers that can take on positive and negative values, we append an additional sign bit and the representation size is given by

$$\mathcal{L}[u \mid \delta] = 1 + \mathcal{B}(|u|; \delta) \quad bits. \tag{5.18}$$

It is important to realize that the complexity (or representation size) of real numbers is specified at a given discretization $\delta$. The same real number $u$ quantized to a smaller interval $\delta$ will have a greater complexity. Therefore, in order to *compare* the complexities of two numbers, they must be expressed at the same level of discretization.

The complexity of a real number $u \geq 0$ having Sig($u$) *significant bits* is $\mathcal{B}(u; 2^{-\lfloor -\log_2 u \rfloor - \mathrm{Sig}(u)})$ bits, where the quantization is chosen to convert the fractional part into an integer. Thus, a number expressed with more significant digits has higher complexity.

### 5.5.5  *Finite Interval*

A finite interval $[t_1, t_2]$ can be encoded by discretizing it to a quantization interval $\delta$, and encoding the resulting discrete set. The representation size of $t \in [t_1, t_2]$ at a discretization $\delta$ is given by

$$\mathcal{L}[t \mid t \in [t_1, t_2], \delta] = \lceil \log_2(\lfloor \frac{|t_2 - t_1|}{\delta} \rfloor + 1) \rceil$$
$$\approx \log_2(|t_2 - t_1| + 1) - \log_2 \delta \quad bits. \tag{5.19}$$

# Multisensor Fusion and Model Selection Framework

# Chapter 6

# Environment and Sensor Models

We discussed general modeling issues in multisensor data fusion in Section 3.2. A tactile–visual multisensor fusion problem was introduced in Section 1.2 on page 2 (Figure 1.1). In this chapter, abstract models for a class of model-based multisensor fusion problems are introduced. The tactile–visual fusion problem is used as example of this class of problems, to illustrate the abstract models.

## 6.1 Environment and Sensor Models

An abstract model of multisensor observations is shown in Figure 6.1, which shows $S$ sensors observing an unknown environment. The underlying environment is described by an **environment model**, $q_E = \Xi(\theta)$, where $\Xi$ denotes the model structure, and $\theta$ denotes a particular parameter instantiation of the environment model. The environment model parameters are drawn from a $D$-dimensional parameter set $\theta \in \Theta \subset \Re^D$, which can be discretely enumerated. In the tactile–visual fusion problem, an environment model is a polyhedral shape structure and the associated six pose parameters.

Each sensor observes some particular aspect of the underlying environment model, denoted by a set of $M$ *model features*, $\mathbf{Y} = \{\mathbf{y}\}_{1,M}$, and produces a set of $N$ *data features*, denoted by $\mathbf{Z} = \{\mathbf{z}\}_{1,N}$. The mapping from an environment model to the model features is denoted by a *model feature extractor*, $\mathcal{F}$, where $\{\mathbf{y}\} = \mathcal{F}(q_E; \alpha)$ is a function of the environment model and the *sensor calibration* $\alpha$. Thus, for the tactile sensor, this mapping extracts the object faces from the instantiated shape structure, while

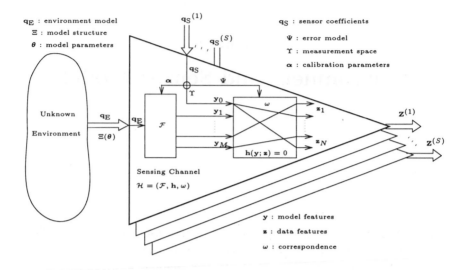

Fig. 6.1   Environment and sensor models. Observed data features $z_i$ arise from corresponding model features $y_{\omega(i)}$ according to a sensor constraint equation $h(y_{\omega(i)}; z_i) = 0$ and observation errors $\Psi$. The model features $\{y\}_{1,M}$ describe some aspect of the parameterized environment model $q_E = \Xi(\theta)$ for the sensor calibration $\alpha$, while $y_0$ denotes outliers in the sensor observation space $\Upsilon$.

for a vision sensor it extracts the visible object vertices, which depend on calibration of the camera viewpoint.

Ideally, a data feature, $z$, may be related to a *model feature*, $y$, by a general constraint $h$:

$$h(y; z) = 0. \tag{6.1}$$

In the tactile–visual fusion problem, the contact data features obtained from a tactile sensor can be related to the object face model features by a planar constraint equation.  The image vertex data features can be related to the object vertex features according to a perspective projection constraint under a pinhole camera model.

In reality, observed data features differ from their ideal values given by the constraint equation, due to errors in sensor measurements. The sensor uncertainty models, denoted by $\Psi$, describe the observation errors, and are assumed to be available *a priori* from previous experimentation with the sensors.

Some observed data features may not be related to the underlying environment model; these are referred to as *outliers* or *unmodeled* data features, in contrast to the *modeled* data features which are related to the environment model. The *special symbol* $y_0$ is used to denote the entire measurement space $\Upsilon$. Outliers or unmodeled data features may lie anywhere in the sensor measurement space, and therefore correspond to the special symbol $y_0$.

The association between the observed data features and the model features or $y_0$, is given by a *correspondence*, $\omega$, that maps data feature indices to model feature indices, or the special index 0 (for outliers):

$$\omega : \{1, \dots, N\} \longrightarrow \{0, 1, \dots, M\}.$$

The correspondence is often unknown, and may be *many-to-one*, as is the case for a tactile sensor where several contact points may correspond to the same object face; or *one-to-one* as is the case for a vision sensor where at most one image vertex point may correspond to an object vertex*.

In practice, raw sensor data features expressed in *local* measurement coordinates, must be converted to a reference frame that can be related to similar reference frames of other sensors. This mapping is known as *data conditioning*, and often depends on the sensor calibration. Sometimes, an instantiated environment model in a global reference frame may be "mapped" via the model features into the local sensor reference frames. Using a combination of these two approaches, the sensor reference frames are aligned, or *registered*, before fusion can take place. The sensor registration usually depends heavily on reliable sensor calibration $\alpha$, which is assumed to be available *a priori*.

The model feature extractor $\mathcal{F}$, the sensor constraint $\mathbf{h}$, and the correspondence $\omega$ are collectively referred to as a *sensing channel* $\mathcal{H} = (\mathcal{F}, \mathbf{h}, \omega)$; they define a sensor structure which has the same form for all sensors of a given type. However, the uncertainty, the measurement range, and the calibration parameters, differ between different sensors of the same type and must be determined separately for each sensor; they are collectively referred to as the *sensor coefficients* $\mathbf{q}_s = (\Psi, \Upsilon, \alpha)$. The sensing channel and the sensor coefficients together constitute a **sensor model**, $\{\mathbf{q}_s, \mathcal{H}\}$.

A complete model, $\mathbf{q}$, of the multisensor observations is given by specify-

---

*In case of occlusions and/or accidental alignments of object vertices, we can arbitrarily choose any one of the possible object vertices.

Table 6.1   Notation used to describe the environment and sensor models.

| Symbol | Meaning |
|--------|---------|
| $\theta$ | environment model parameters |
| $\Theta$ | space of legitimate environment model parameters |
| $\Xi$ | environment model structure |
| $\mathbf{q}_E$ | environment model, $\Xi(\theta)$ |
| $\mathbf{Q}_E$ | library of environment models |
| $\Psi$ | sensor uncertainty model describing observation errors |
| $\Upsilon$ | sensor measurement space |
| $\alpha$ | sensor calibration |
| $\mathbf{q}_S$ | sensor coefficients, $(\Psi, \Upsilon, \alpha)$ |
| $\mathbf{z}$ | data features |
| $\mathbf{y}$ | model features |
| $\omega$ | correspondence mapping data features to model features |
| $\mathbf{h}$ | constraint relating data and model features, $\mathbf{h}(\mathbf{y}; \mathbf{z}) = 0$ |
| $\mathcal{F}$ | model feature extraction from the environment model, $\{\mathbf{y}\} = \mathcal{F}(\mathbf{q}_E; \alpha)$ |
| $\mathcal{H}$ | sensing channel, $(\mathcal{F}, \mathbf{h}, \omega)$ |
| $\mathbf{q}$ | multisensor observation model, $(\mathbf{q}_E, \{\mathbf{q}_S, \mathcal{H}\})$ |

ing the environment model $\mathbf{q}_E$, and the multiple sensor models $\{\mathbf{q}_S, \mathcal{H}\}_{1,S}$:

$$\mathbf{q} = (\mathbf{q}_E, \{\mathbf{q}_S, \mathcal{H}\}_{1,S}). \qquad (6.2)$$

The notation used to describe the abstract multisensor fusion problem is summarized in Table 6.1.

## 6.2   Correspondence Models

The data correspondence $\omega$ is diagrammatically shown in Figure 6.2. The special symbol $\mathbf{y}_0$ denotes outliers. We use the following nomenclature:

$$\omega(i) = \begin{cases} 0 & \mathbf{z}_i \text{ is "unmodeled" i.e., an outlier.} \\ j & \mathbf{z}_i \text{ is "modeled" and corresponds to } \mathbf{y}_j. \end{cases}$$

Thus, the set of data features for which $\omega(i) \neq 0$ is referred to as "modeled" data features. The correspondence may be an onto mapping—there may be

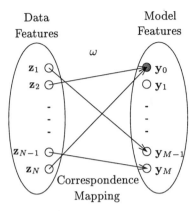

Fig. 6.2   The data correspondence is a discrete mapping from the observed data feature indices to the model feature indices or the special symbol $\mathbf{y}_0$ (for outliers).

model features to which no data feature corresponds. Such model features are referred to as "missing" model features. The set of all allowed or *legal* correspondences is denoted by $\Omega$; $\omega \in \Omega$.

### 6.2.1   *Many-to-One Correspondence*

The case of *many-to-one* correspondence occurs when there is no restriction on the mapping $\omega$. Any number of data features can correspond to a model feature or $\mathbf{y}_0$. The space of legal correspondences, $\Omega$, includes all possible correspondences, and

$$|\Omega| = (M + 1)^N$$

i.e. there are $|\Omega|$ allowed correspondence mappings.

### 6.2.2   *One-to-One Correspondence*

The case of *one-to-one correspondence* occurs when a one-to-one restriction is placed on modeled data features—at most one data feature may correspond to a model feature $\mathbf{y}_1 \ldots \mathbf{y}_M$. However, there is no restriction on the number of outliers; any number of data features may correspond to $\mathbf{y}_0$.

The space of legal correspondences, $\Omega$, is the set of all one-to-one map-

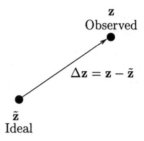

Fig. 6.3   An observed data feature **z** differs from its ideal value **z̃**.

pings from the subset of modeled data features to model features. Thus

$$|\Omega| = \sum_{k=0}^{\min(M,N)} \left( \begin{array}{c} M \\ k \end{array} \right) \left( \begin{array}{c} N \\ k \end{array} \right) k!$$

where $k$ denotes the number of modeled data features.

## 6.3   Uncertainty Models

An observed data feature **z** differs from its ideal value **z̃**, where the tilde, ˜, denotes an *ideal* or predicted data feature value based on an environment model. The sensing uncertainty introduces the observation error $\Delta\mathbf{z} = \mathbf{z}-\tilde{\mathbf{z}}$, as shown in Figure 6.3. Sensor uncertainty or observation errors may arise from several sources, some of which are listed below.

- **Measurement Noise:** This is the most common type of uncertainty, inherent in the physics of the measurement process.
- **Gross Errors:** Gross errors are due to spurious data or outliers in the measurements.
- **Quantization Errors:** Discretization of the data leads to a quantization errors in the observed value. These are bounded and depend on the quantization interval.
- **Numerical Errors:** Data is processed using digital computers that operate on finite length bit strings. The finite precision of a digital computer gives rise to numerical errors.
- **Control Errors:** Control errors are due to the inaccuracies in controlling sensor parameters. For example, tactile sensors mounted

Fig. 6.4   The sensor measurement space is uniformly subdivided into uncertainty regions of specified shape and volume.

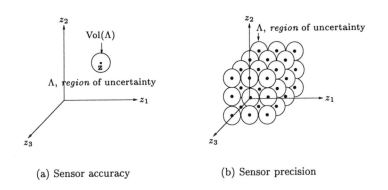

(a) Sensor accuracy                    (b) Sensor precision

on a hand/arm system will have errors due to the uncertainty in controlling the hand position.

Sensor uncertainty can be mathematically described in a number of ways, discussed next.

### 6.3.1   *Sensor Accuracy and Precision*

The **sensor accuracy** is defined as the finite *region of uncertainty*, $\Lambda$, around a measured value as shown in Figure 6.5(a). The uncertainty region is characterized by some *shape* and *volume*, $\mathrm{Vol}(\Lambda)$. The true value of the data feature may lie anywhere in accuracy region around the observed value. The sensor **measurement space**, $\Upsilon$, is a finite subset of $\Re^n$. It is assumed to be uniformly *tiled* into uncertainty regions of specified shape and volume as shown in Figure 6.5(b). We use the phrase "higher accuracy" to mean a smaller value of $\mathrm{Vol}(\Lambda)$ i.e., "lower uncertainty", and vice-versa.

The representation size or the *smallest* number of bits needed to encode the observation error $\Delta \mathbf{z}$, may be found by encoding the observed data feature $\mathbf{z}$ relative to the ideal data feature $\tilde{\mathbf{z}}$. A **radial-polar encoding scheme** encodes an observed data feature $\mathbf{z}$ using *polar coordinates* centered at $\tilde{\mathbf{z}}$, as shown in Figure 6.5. A **spiral encoding scheme** encodes an observed data feature by counting spirally outwards from the ideal data

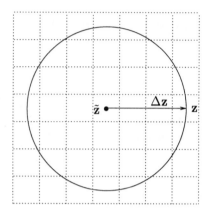

Fig. 6.5  In a *radial-polar* encoding scheme, the observed data feature is encoded by specifying the polar coordinates of **z** relative to the ideal data feature $\tilde{\mathbf{z}}$.

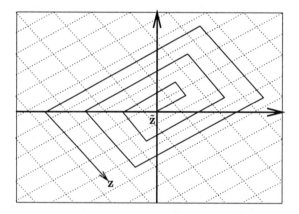

Fig. 6.6  In a *spiral encoding scheme*, an observed data feature **z** is specified relative to $\tilde{\mathbf{z}}$, by counting spirally outwards from $\tilde{\mathbf{z}}$ and encoding the integer index obtained for **z**.

feature $\tilde{\mathbf{z}}$, until the observed data feature **z** is reached, as shown in Figure 6.6, and then encoding this number using an *integer coding scheme* $\mathcal{I}(\cdot)$ (Section 5.5 on page 66). The spiral count is given by the ratio $\lfloor \mathrm{Vol}(\Delta\mathbf{z})/\mathrm{Vol}(\Lambda) \rfloor$, where $\mathrm{Vol}(\Delta\mathbf{z})$ is the volume of the region spanned by

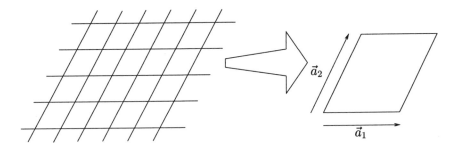

Fig. 6.7 Parallelepiped uncertainty regions in two-dimensional measurement space.

the spiral. Therefore, the representation size is given by

$$\mathcal{L}[\Delta \mathbf{z} \,|\, \tilde{\mathbf{z}}] = \mathcal{I}(\lfloor \frac{\text{Vol}(\Delta \mathbf{z})}{\text{Vol}(\Lambda)} \rfloor) \approx \log_2 (\frac{\text{Vol}(\Delta \mathbf{z})}{\text{Vol}(\Lambda)} + 1) \qquad (6.3)$$

bits.

The **sensor precision** is defined as the *number of bits* needed to address any point in the measurement space as shown in Figure 6.5(b). Given a measurement space, $\Upsilon$, the sensor precision, $\lambda$, is given by

$$\lambda \overset{\Delta}{=} \lceil \log_2 (\lfloor \frac{\text{Vol}(\Upsilon)}{\text{Vol}(\Lambda)} \rfloor + 1) \rceil \approx \log_2 (\frac{\text{Vol}(\Upsilon)}{\text{Vol}(\Lambda)} + 1) \qquad (6.4)$$

bits.

### 6.3.1.1 *Ellipsoidal Regions*

While there are many possible choices for the shape of the uncertainty region, $\Lambda$, let us consider *parallelepiped* shaped uncertainty regions, as shown in Figure 6.7. Such regions define a *uniform grid* in the sensor measurement space, and are commonly used for sampling sensor measurements in multi-dimensional digital signal processing [49], and in building fault-tolerant sensors [133].

Given an $n$-dimensional measurement space, a parallelepiped uncertainty region is specified by $n$ vectors $\Lambda = (\vec{a}_1, \dots, \vec{a}_n)$, which specify the $n$ sides of the parallelepiped. For ease of mathematical manipulation, let us approximate these parallelepiped shaped regions by *ellipsoidal* shaped regions. As shown in Figure 6.8, a parallelepiped region is approximated by the smallest *enscribing* ellipsoid. An ellipsoidal uncertainty region em-

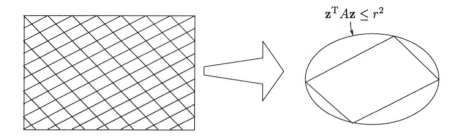

Fig. 6.8   Ellipsoidal uncertainty regions in two-dimensional measurement space.

bedded in $n$-dimensional measurement space is described by a quadratic form:

$$\mathbf{z}^{\mathrm{T}} A \mathbf{z} \le r^2,$$

where $A$ is a $n \times n$ symmetric matrix specifying the shape and size of the ellipsoid, and $r$ is the radius of the ellipsoid. Its volume is given by Rissanen [166]

$$\mathrm{Vol}(\Lambda) = \left(\frac{4r^2}{n}\right)^{n/2} \frac{1}{|A|^{1/2}}. \tag{6.5}$$

The radius is typically chosen to be $r = 1$, since we can always normalize by replacing $A$ by $A/r^2$.

The observation error representation size can be easily calculated, using a spiral encoding scheme (Figure 6.6 on page 82). The counting spiral spans an ellipsoid of the same shape as the uncertainty region, but larger radius:

$$\mathbf{z}^{\mathrm{T}} A \mathbf{z} \le \Delta \mathbf{z}^{\mathrm{T}} A \Delta \mathbf{z}.$$

Its volume is given by

$$\mathrm{Vol}(\Delta \mathbf{z}) = \left(\frac{4\Delta \mathbf{z}^{\mathrm{T}} A \Delta \mathbf{z}}{n}\right)^{n/2} \frac{1}{|A|^{1/2}}.$$

Therefore, the representation size of the observation error, given an ideal data feature $\tilde{\mathbf{z}}$, is given by (Equation 6.3 on the page before):

$$\mathcal{L}[\Delta \mathbf{z} \,|\, \tilde{\mathbf{z}}] = \mathcal{I}(\lfloor \frac{\mathrm{Vol}(\mathbf{z}^{\mathrm{T}} A \mathbf{z} \le \Delta \mathbf{z}^{\mathrm{T}} A \Delta \mathbf{z})}{\mathrm{Vol}(\mathbf{z}^{\mathrm{T}} A \mathbf{z} \le r^2)} \rfloor) \approx \mathcal{I}(\left[\Delta \mathbf{z}^{\mathrm{T}} \frac{A}{r^2} \Delta \mathbf{z}\right]^{n/2}), \tag{6.6}$$

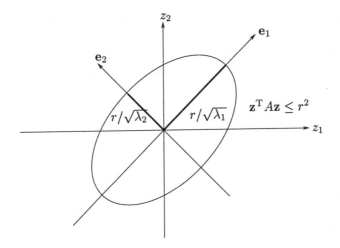

Fig. 6.9   A two-dimensional ellipsoidal uncertainty region. The eigenvectors $e_1$, $e_2$ and the eigenvalues $\lambda_1$, $\lambda_2$, of matrix $A$, respectively give the directions and the radial lengths of the principal axes.

and may be approximated as $\log_2((\Delta z^T \frac{A}{r^2} \Delta z)^{n/2} + 1)$ (Equation 5.12 on page 68). Note that this expression essentially encodes the $n$-th power of the *Mahalanobis norm*, $\|\Delta z\|_{A/r^2}$, where $\|x\|_M \triangleq (x^T M x)^{1/2}$.

The measurement space is described by a larger ellipsoid of the same shape as the uncertainty region, but radius $R$:

$$z^T A z \leq R^2.$$

Therefore, the sensor precision is given by:

$$\lambda \approx \log_2\left(\left[\frac{R}{r}\right]^n + 1\right), \tag{6.7}$$

using Equation (6.4).

### Example 6.1   1-D Measurement Space

For a 1-dimensional measurement space with quantization $\delta$, and $r = 1$, we have $A = 1/\delta^2$, The error representation size is given by $\mathcal{L}[\Delta z \,|\, \bar{z}] \approx \log_2(|\Delta z|/\delta + 1)$.

### Example 6.2   2-D Measurement Space

Figure 6.9 shows a 2-D ellipsoidal uncertainty region, $z^T A z \leq r^2$, where the *eigenvectors* $e_1$ and $e_2$ of the matrix $A$ give the directions of the principal axes, while the *eigenvalues* $\lambda_1$ and $\lambda_2$ specify their radial lengths.

As the the radius $r$ increases, or as the eigenvalues $\lambda$ decrease, the volume of the uncertainty region increases, and we say that the accuracy is lower, and vice-versa.

The *similarity transformation* $P$ describes the principal axes coordinate frame $e_1, e_2$ in the $z_1, z_2$ coordinate frame, and is defined as:

$$P = \left[ \begin{array}{cc} \sqrt{\lambda_1} & 0 \\ 0 & \sqrt{\lambda_2} \end{array} \right] [e_1 \quad e_2].$$

Since, by definition, $Ae = \lambda e$, therefore we have $A = P^T P$.

We can simulate errors described by an ellipsoidal uncertainty, by generating errors in the principal coordinate frame along each principal direction (we often use *uniform* i.i.d. random variables), and then using the similarity transformation $P$ to transform these to the $z_1, z_2$ coordinate frame; thus $\Delta z = P[e_1 \quad e_2]^T$ where $[e_1 \quad e_2]^T$ is an error vector in the principal axes coordinate frame. This procedure results in an elliptical sampling of the original measurement space.

### 6.3.2   *Probability Distributions*

In a probabilistic uncertainty model , observation errors are described by a conditional probability distribution,

$$p_{z \mid \tilde{z}}(z \mid \tilde{z}) = p(\Delta z \mid \tilde{z}),$$

and are often assumed to be independent of the ideal data feature value $\tilde{z}$; thus

$$p_{z \mid \tilde{z}}(z \mid \tilde{z}) = p(\Delta z).$$

The representation size of the observation errors is the *smallest* number of bits needed to encode the error, $\Delta z$, and is given by (Section 5.5 on page 66)

$$\mathcal{L}[\Delta z \mid \tilde{z}] = -\log_2 p(\Delta z \mid \tilde{z}) - \sum_{k=1}^{n} \log_2 \delta_k \quad bits \qquad (6.8)$$

where $\delta_k$ is the quantization interval along the $z_i$ axis of the $n$-dimensional measurement space. This expression uses an optimal encoding of a prob-

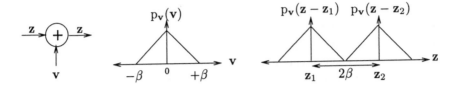

Fig. 6.10 Uncertainty region induced by a 1-dimensional signal contaminated by independent additive measurement noise. Two signal values $z_1, z_2$ can be distinguished unambiguously in the observed signal $z$, if they are separated by a step size of $2\beta$, where $2\beta$ is the finite support region of the noise density.

ability distribution, and therefore gives the best data compression on an average.

In practice, a sensor operates in some small sub-space of all possible $n$-dimensional data features, $\Re^n$. This sensor **measurement space** is formally defined by specifying a probability distribution, $p_0(z)$, on the observed data features. The number of bits need to specify a point $z$ in this measurement space defines the *probabilistic sensor precision*:

$$\lambda = -\log_2 p_0(z) - \sum_{k=1}^{n} \log_2 \delta_k. \qquad (6.9)$$

We can define the **average sensor precision**, $\bar{\lambda}$, as the *average* number of bits required to specify any point in the measurement space,

$$\bar{\lambda} \triangleq E\{-\log_2 p_0(z)\} - \sum_{k=1}^{n} \log_2 \delta_k \qquad (6.10)$$

where $E\{\cdot\}$ denotes expected value [151].

### 6.3.2.1 *Probability Regions*

We can also define the notion of sensor accuracy for probabilistic uncertainty models, by considering the uncertainty region induced by a probability distribution.

Consider a one-dimensional sensor signal $z$ contaminated by independent additive zero mean measurement noise $v$, as shown in Figure 6.10. The probability density of the observed signal $z = z + v$ is is given by the convolution of the the densities $p_z(z)$ and $p_v(v)$ [151, page 136]: $p_z(z) = p_z(z) \otimes p_v(v)$. Suppose two signal samples $z_1, z_2$ are observed, and assume $p_z(z) = \delta(z - z_i), i = 1, 2$. The probability density for the observed signal from the first sample is $p_z(z) = \delta(z - z_1) \otimes p_v(v) = p_v(z - z_1)$, while for the second sample it is $p_z(z) = p_v(z - z_2)$. For a zero mean symmetric noise density having a *finite support* region of width $2\beta$ as shown in Figure 6.10, the two samples can be distinguished unambiguously if their separation is at least $2\beta$, i.e. $|z_1 - z_2| \geq 2\beta$. We define this finite support region of $p_v(v)$ to be the uncertainty region induced by the the additive measurement noise $v$.

In general, for multidimensional data features, we define the $\epsilon$-*uncertainty region*, $\Lambda_\epsilon$, as the *minimum volume* region such that the probability of an observed data feature lying outside it is less than $\epsilon$:

$$\text{Vol}(\Lambda_\epsilon) \stackrel{\Delta}{=} \min\{\text{Vol}(\Lambda) : P_v(v \notin \Lambda) \leq \epsilon\} = \int_{p(\Delta z) > \epsilon} d\Delta z. \qquad (6.11)$$

The shape of the uncertainty region depends on the region of support for the noise density $p_v(v)$. For *Normal* distributions [151], this is an ellipsoidal region, described by the inverse noise covariance matrix, $A \propto \Sigma^{-1}$. The *Tchebycheff Inequality* [151, page 113] may be useful in defining ellipsoidal uncertainty regions for other noise distributions.

For probabilistic uncertainty models, we can define the $\epsilon$-**sensor accuracy** to be the $\epsilon$-uncertainty region for some suitably chosen values of $\epsilon$ (say, 1%), given a distribution, $p(\Delta z)$, on the observation errors.

### 6.3.3 *Practical Determination of Sensor Accuracy and Precision*

Sensor accuracy and precision can be determined in several ways, in practice.

(1) **Data Quantization:** Some sensors are characterized by an inherent quantization of data features, and this quantization window or "step size" may be used to define the uncertainty region. For instance, spatial image data obtained from a camera is inherently quantized by the *pixel size and shape*, which defines a parallelepiped uncertainty region. The measurement space is defined by the sensor's operating ranges of observable data features; in the case of a camera, this is the image size.

(2) **Data Covariance:** The inverse of the data covariance matrix, $\hat{\Sigma}$,

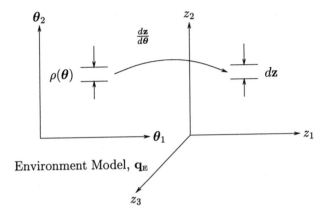

**Fig. 6.11** Sensor resolution is defined as the smallest change in the environment model parameters which produces a detectable change in the observed data features.

computed from observed data, may be used as an approximation of an ellipsoidal accuracy region: $A \propto \hat{\Sigma}^{-1}$. The measurement space may be approximated by the sensor's operating range of measurable feature values.

## 6.4 Sensor Resolution

We define **sensor resolution**, $\rho(\boldsymbol{\theta})$, as the smallest *change* in the environment model parameters which produces a *detectable* change in the observed data features, for a given environment model, $\Xi(\boldsymbol{\theta})$. As illustrated in Figure 6.11, it relates changes in the environment model to the changes in observed data, and therefore depends on the constraints $\mathbf{h}$ relating observed data features to the model features, the model feature extractor $\mathcal{F}$, and the environment model parameters $\boldsymbol{\theta}$, for a given choice model structure, parameters and correspondences (interpretation).

Given an environment model structure and parameters, the sensor resolution is defined as:

$$\rho(\boldsymbol{\theta}) = \frac{d\mathbf{z}}{d\boldsymbol{\theta}}. \tag{6.12}$$

A larger value of $\rho(\boldsymbol{\theta})$ will produce a greater change in the observed data features, for the same change in the environment model parameters.

Often, the sensor resolution is specified as the change in model parameters, that would produce a change equal to the sensor accuracy, $\Lambda$; thus

$$\delta\boldsymbol{\theta} = \frac{\Lambda}{\rho(\boldsymbol{\theta})}, \tag{6.13}$$

and is referred to as **parameter resolution**. Thus, for a given sensor resolution, $\rho(\boldsymbol{\theta})$, the higher the accuracy (i.e., smaller $\Lambda$), the finer the parameter resolution $\delta\boldsymbol{\theta}$.

In general, the parameter resolution, $\delta\boldsymbol{\theta}$ is a function of the number of data features $N^{(s)}$ from each sensor, and the sensor accuracy, $\Lambda^{(s)}$, where $s = 1,\ldots,S$. The parameter resolution, $\delta\boldsymbol{\theta}$ decreases (improves) as the number of data features $N$ increases, or as the sensor accuracy, $\Lambda$ decreases (higher accuracy).

Intuitively, sensor resolution is the ability of a sensor to *detect change* in the environment. Suppose a range sensor measures the distance of an object 10m away within $\pm$10cm. Then, the sensor accuracy is 20cm. Suppose the sensor can detect object movements of 5cm—i.e., moving the object by 5cm changes the sensor reading by more than 20cm. Then, the sensor resolution is $20/5 = 4$cm/cm. Detection of smaller object movements (i.e., a finer parameter resolution) can be achieved by (a) using a higher accuracy sensor (i.e., decreasing $\Lambda$), or (b) by increasing the sensor resolution $\rho(\boldsymbol{\theta})$.

# Chapter 7

# Minimal Representation Multisensor Fusion and Model Selection

In multisensor fusion and model selection problems, the data from several different sensors is collected. A model which explains the data well provides a good interpretation of the underlying environment. The selection of a good candidate model, the measure of consistency among disparate sensor types, and the treatment of noisy data by rejection of outliers are problems of particular importance. In this chapter we express multisensor fusion problem in a form suitable for use with the minimal representation size criterion, and develop a framework which addresses these model selection issues.

## 7.1 Multisensor Fusion and Model Selection

In terms of the abstract environment and sensor models introduced in Section 6.1, the multisensor fusion and model selection problem (Figure 7.1) is to select the environment structure and parameters, $\hat{q}_E \in Q_E$, and the data correspondences, $\{\omega \in \Omega\}_{1,S}$, given an environment model library, $Q_E$, the sensor coefficients, $q_S$, the constraint relations $\{h\}_{1,S}$, and the model feature extractors $\{\mathcal{F}\}_{1,S}$ for each sensor. The environment model, $q_E = \Xi(\theta)$, and the data correspondences, $\{\omega\}_{1,S}$, define an *interpretation* of the environment.

The block diagram of a multisensor fusion and model selection system is shown in Figure 7.2. For each sensor, the calibration and uncertainty model are assumed to be known *a priori*. These may have been determined analytically, or experimentally, as a result of previous estimation procedures. The raw sensor data features are often be conditioned to a form suitable

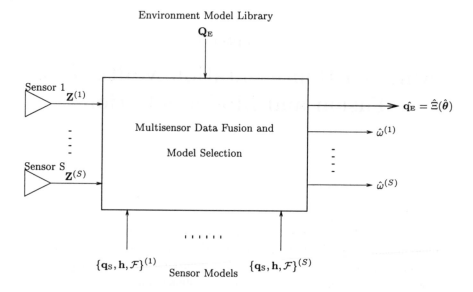

Fig. 7.1   The multisensor fusion and model selection problem.

for use in the sensor constraint equation. An environment model library $\mathbf{Q}_E$ provides potential explanations of the unknown environment. Given an environment model $\mathbf{q}_E = \Xi(\boldsymbol{\theta})$ and the model feature extraction procedure $\mathcal{F}$, the model features are computed for each sensor. The conditioned data features correspond to extracted model features, via the sensor constraint equation (see Figure 6.1 on page 76).

In an ideal setting, these constraints would be satisfied exactly by the observed data when the correct interpretation is selected. Given enough number of such constraints, it is possible to select the correct underlying interpretation which resulted in the observed multisensor data. The output of the system is the selected interpretation which "best" explains the observed multisensor data, and is specified in terms of the environment model $\hat{\mathbf{q}}_E = \hat{\Xi}(\hat{\boldsymbol{\theta}})$, and the correspondences $\hat{\omega}$ for each sensor.

We formulate this multisensor fusion and model selection problem using a minimal representation size criterion. The minimal representation size criterion is used to (a) choose among alternatives, $\mathbf{q}_E \in \mathbf{Q}_E$, from a library of environment models, (b) choose the environment model parameter resolution and data scaling based on the sensor resolution and accuracy

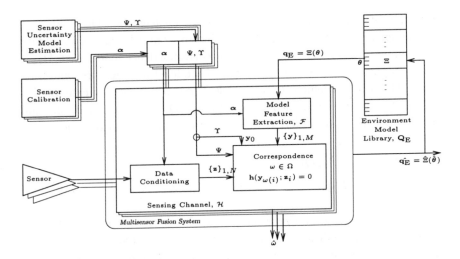

Fig. 7.2 Block diagram of a multisensor fusion and model selection system. The inputs are the sensor coefficients $q_S = (\Psi, \Upsilon, \alpha)$, the environment model library $Q_E$, and the raw sensor data. The outputs are the selected environment model structure and parameters, $\hat{\Xi}(\hat{\theta})$, and the correspondences, $\hat{\omega}$, for each sensor.

of data features from different sensors, (c) choose the subset of observed data features which are modeled features (and are therefore used in the environment model parameter estimation), and (d) choose the correspondence, $\{\omega \in \Omega\}_{1,S}$, for each sensor that maps its data features to model features.

In addition, the minimal representation size criterion may be used for the environment model parameter estimation itself, or alternatively a more traditional statistical estimation method may be used.

## 7.2 Minimal Representation Multisensor Fusion

In this section, the minimal representation size principle is applied to the multisensor fusion and model selection problem. The multisensor data representation size is given by:

$$\mathcal{L}[\mathbf{q}, \mathbf{Z}^{(1)}, \dots, \mathbf{Z}^{(S)} \mid \mathbf{Q}]$$

where $\mathbf{q} \in \mathbf{Q}$ is a selected model chosen from a collection of environment and sensor models (see Equation 6.2 on page 78), and $\mathbf{Z}^{(s)}$ is observed data from the $s$-th sensor, as shown in Figure 6.1 on page 76. The unknowns are the environment model $q_E \in Q_E$ selected from a library of candidate environment models $Q_E$, and the sensor correspondence model $\{\omega \in \Omega\}_{1,S}$ which implicitly includes the selection of modeled data features and outliers. The sensor coefficients, $q_s = (\Psi, \Upsilon, \alpha)$, are known or estimated *a priori*. Given an environment model library $Q_E$, and the set of all legal correspondence models $\{\Omega\}_{1,S}$, we can simplify the above expression as:

$$\mathcal{L}[q, Z^{(1)}, \ldots, Z^{(S)} \mid Q]$$
$$= \mathcal{L}[q_E, \omega^{(1)}, \ldots, \omega^{(S)}, Z^{(1)}, \ldots, Z^{(S)} \mid Q_E, \Omega^{(1)}, \ldots, \Omega^{(S)}],$$

where the representation size is computed relative to the space of all possible environment models in the library, and the space of all legal correspondence models. For convenience we rewrite this as: $\mathcal{L}[q, Z^{(1)}, \ldots, Z^{(S)} \mid Q] = \mathcal{L}[q_E, \omega^{(1)}, \ldots, \omega^{(S)}, Z^{(1)}, \ldots, Z^{(S)} \mid Q_E]$. Simplifying:

$$\mathcal{L}[q, Z^{(1)}, \ldots, Z^{(S)} \mid Q]$$
$$= \mathcal{L}[q_E \mid Q_E] + \mathcal{L}[\omega^{(1)}, \ldots, \omega^{(S)}, Z^{(1)}, \ldots, Z^{(S)} \mid q_E, Q_E],$$

where the "best" interpretation, specified by $q_E \in Q_E$ and $\{\omega \in \Omega\}_{1,S}$, has to be selected.

According to the minimal representation size principle, the best interpretation is given by the choice of $q_E$ and $\{\omega\}_{1,S}$ that minimize the total representation size, $\mathcal{L}[q, Z^{(1)}, \ldots, Z^{(S)} \mid Q]$. The minimal representation criterion for the multisensor fusion and model selection problem becomes,

$$\hat{q_E} = \arg \min_{q_E \in Q_E} \mathcal{L}[q_E \mid Q_E] + \mathcal{L}[Z^{(1)}, \ldots, Z^{(S)} \mid q_E, Q_E] \qquad (7.1)$$

where

$$\mathcal{L}[Z^{(1)}, \ldots, Z^{(S)} \mid q_E, Q_E] \triangleq$$
$$\min_{\{\omega \in \Omega\}_{1,S}} \mathcal{L}[\omega^{(1)}, \ldots, \omega^{(S)}, Z^{(1)}, \ldots, Z^{(S)} \mid q_E, Q_E] \quad (7.2)$$

is the smallest encoding of observed data given the environment model $q_E$, and $\Omega$ is the set of all legal correspondences. The correspondences can be selected using the same criterion

$$(\hat{\omega}^{(1)}, \dots, \hat{\omega}^{(S)}) = \arg\min_{\{\omega \in \Omega\}_{1,S}} \mathcal{L}[\omega^{(1)}, \dots, \omega^{(S)}, \mathbf{Z}^{(1)}, \dots, \mathbf{Z}^{(S)} \mid \hat{\mathbf{q}}_{\mathrm{E}}, \mathbf{Q}_{\mathrm{E}}]$$

(7.3)

where $\hat{\mathbf{q}}_{\mathrm{E}}$ is the selected environment model from Equation (7.1). Note that $\Omega = \Omega(\mathbf{q}_{\mathrm{E}})$, since the number of model features depends on model feature extraction. The search for the minimal representation environment model and correspondences spans the set of all environment model structures in the library in all possible parameter instantiations, and all legal correspondences.

### 7.2.1 *Independent Sensors*

Sensors are defined to be *independent* iff the each sensor's data can be encoded independently of other sensors; i.e.,

$$\mathcal{L}[\omega^{(1)}, \dots, \omega^{(S)}, \mathbf{Z}^{(1)}, \dots, \mathbf{Z}^{(S)} \mid \mathbf{q}_{\mathrm{E}}, \mathbf{Q}_{\mathrm{E}}] = \sum_{s=1}^{S} \mathcal{L}[\omega^{(s)}, \mathbf{Z}^{(s)} \mid \mathbf{q}_{\mathrm{E}}, \mathbf{Q}_{\mathrm{E}}] \quad (7.4)$$

holds for *all* legal correspondences and all possible observed data. In other words, joint encoding of independent multisensor data does not give any savings in codelength over separately encoding each sensor's data and adding up the codelengths. For sensors described by probabilistic uncertainty models, this requirement is equivalent to having independent probability distributions.

For independent sensors, Equation (7.1) simplifies to

$$\hat{\mathbf{q}}_{\mathrm{E}} = \arg\min_{\mathbf{q}_{\mathrm{E}} \in \mathbf{Q}_{\mathrm{E}}} \mathcal{L}[\mathbf{q}_{\mathrm{E}} \mid \mathbf{Q}_{\mathrm{E}}] + \sum_{s=1}^{S} \mathcal{L}[\mathbf{Z}^{(s)} \mid \mathbf{q}_{\mathrm{E}}, \mathbf{Q}_{\mathrm{E}}]$$

(7.5)

where

$$\mathcal{L}[\mathbf{Z} \mid \mathbf{q}_{\mathrm{E}}, \mathbf{Q}_{\mathrm{E}}] \triangleq \min_{\omega \in \Omega} \mathcal{L}[\omega, \mathbf{Z} \mid \mathbf{q}_{\mathrm{E}}, \mathbf{Q}_{\mathrm{E}}], \quad (7.6)$$

and Equation (7.3) gets decoupled into $S$ independent minimizations:

$$\hat{\omega} = \arg\min_{\omega \in \Omega} \mathcal{L}[\omega, \mathbf{Z} \mid \hat{\mathbf{q}}_{\mathrm{E}}, \mathbf{Q}_{\mathrm{E}}].$$

(7.7)

In the rest of this book, we assume that sensors are independent, and use this form of the minimal representation size criterion. In practice, the sensor dependencies are often unknown and difficult to characterize. Furthermore, the independence assumption simplifies the correspondence search space into several *disjoint* search spaces (Equation 7.7).

The representation size of the correspondence and the observed data features for a sensor, given by $\mathcal{L}[\omega, \mathbf{Z} \mid \mathbf{q}_E, \mathbf{Q}_E]$, can be written as:

$$\mathcal{L}[\omega, \mathbf{Z} \mid \mathbf{q}_E, \mathbf{Q}_E] = \mathcal{L}[\mathcal{A} \mid \mathbf{q}_E, \mathbf{Q}_E] + \mathcal{L}[\omega, \mathbf{Z} \mid \mathcal{A}, \mathbf{q}_E, \mathbf{Q}_E],$$

where $\mathcal{A}$ is an algorithm used to encode the data and correspondences, given an environment model selected from the environment model library. The representation size of the encoding algorithm is given by $\mathcal{L}[\mathcal{A} \mid \mathbf{q}_E, \mathbf{Q}_E]$, while the representation size of the correspondence and observed data, obtained using this encoding algorithm, is given by $\mathcal{L}[\omega, \mathbf{Z} \mid \mathcal{A}, \mathbf{q}_E, \mathbf{Q}_E]$.

### 7.2.1.1   *Independent Data Features*

Observed data features are defined to be *independent* iff each data feature can be encoded independently of other data features; i.e.,

$$\mathcal{L}[\omega, \mathbf{Z} \mid \mathcal{A}, \mathbf{q}_E, \mathbf{Q}_E] = \sum_{i=1}^{N} \mathcal{L}[\omega, \mathbf{z}_i \mid \mathcal{A}, \mathbf{q}_E, \mathbf{Q}_E] \qquad (7.8)$$

holds for *all* legal correspondences and all possible observed data features. In other words, joint encoding of independent observed data features from a sensor does not give any savings in codelength, over separately encoding each observed data feature and adding up the codelengths. For sensors described by probabilistic uncertainty models, this requirement is equivalent to having independent probability distributions.

In the rest of this book, we assume that the observed data features are independent, and can be encoded separately.

**Mixture Encoding** When the the correspondence $\omega$, and an observed data feature, $\mathbf{z}$, are encoded *jointly*,

$$\mathcal{L}[\omega, \mathbf{z} \mid \mathcal{A}, \mathbf{q}_E, \mathbf{Q}_E]$$

as in Equation (7.8), we refer to this form as the **mixture** encoding.

**Two-Part Encoding** When the correspondence and the data features are encoded *separately*,

$$\mathcal{L}[\omega, \mathbf{Z} \,|\, \mathcal{A}, \mathbf{q}_{\mathrm{E}}, \mathbf{Q}_{\mathrm{E}}] = \mathcal{L}[\omega \,|\, \mathcal{A}, \mathbf{q}_{\mathrm{E}}, \mathbf{Q}_{\mathrm{E}}] + \mathcal{L}[\mathbf{Z} \,|\, \omega, \mathcal{A}, \mathbf{q}_{\mathrm{E}}, \mathbf{Q}_{\mathrm{E}}],$$

we refer to this form as **two-part** encoding. For *independent* data features,

$$\mathcal{L}[\omega, \mathbf{Z} \,|\, \mathcal{A}, \mathbf{q}_{\mathrm{E}}, \mathbf{Q}_{\mathrm{E}}] = \mathcal{L}[\omega \,|\, \mathcal{A}, \mathbf{q}_{\mathrm{E}}, \mathbf{Q}_{\mathrm{E}}] + \sum_{i=1}^{N} \mathcal{L}[\mathbf{z}_i \,|\, \omega, \mathcal{A}, \mathbf{q}_{\mathrm{E}}, \mathbf{Q}_{\mathrm{E}}], \tag{7.9}$$

where each data feature is encoded independently. Note that,

$$\mathcal{L}[\mathbf{z}_i \,|\, \omega, \mathcal{A}, \mathbf{q}_{\mathrm{E}}, \mathbf{Q}_{\mathrm{E}}] \equiv \mathcal{L}[\mathbf{z}_i \,|\, \mathbf{y}_{\omega(i)}],$$

denotes the representation size of data feature $\mathbf{z}_i$, given that it corresponds to the model feature $\mathbf{y}_{\omega(i)}$.

**Known Correspondence** In the special case of a *known correspondence*,

$$\mathcal{L}[\omega, \mathbf{Z} \,|\, \mathbf{q}_{\mathrm{E}}, \mathbf{Q}_{\mathrm{E}}] = \sum_{i=1}^{N} \mathcal{L}[\mathbf{z}_i \,|\, \mathbf{y}_{\omega(i)}],$$

where only the data features need to be encoded, since the correspondence $\omega$ is already known.

## 7.3 Multisensor Fusion Framework

Assuming independent sensors and data features, the total multisensor representation size is obtained by adding the data and correspondence representation size for each sensor, and the common environment model representation size:

$$\mathcal{L}[\mathbf{q}, \mathbf{Z}^{(1)}, \dots, \mathbf{Z}^{(S)} \,|\, \mathbf{Q}] = \mathcal{L}[\mathbf{q}_{\mathrm{E}} \,|\, \mathbf{Q}_{\mathrm{E}}] +$$

$$\sum_{s=1}^{S} \left\{ \mathcal{L}[\mathcal{A}^{(s)} \,|\, \mathbf{q}_{\mathrm{E}}, \mathbf{Q}_{\mathrm{E}}] + \sum_{i=1}^{N^{(s)}} \mathcal{L}[\omega^{(s)}, \mathbf{z}_i{}^{(s)} \,|\, \mathcal{A}^{(s)}, \mathbf{q}_{\mathrm{E}}, \mathbf{Q}_{\mathrm{E}}] \right\}. \tag{7.10}$$

The major components of the total representation size, the environment model representation size, and the sensor representation size, are shown in

Figure 7.3. In the minimal representation size framework, the environment model $\mathbf{q}_E = \Xi(\boldsymbol{\theta})$ and sensor correspondences $\omega^{(1)}, \ldots, \omega^{(S)}$ are selected to minimize the total representation size.

In the rest of this section, we develop encoding schemes for environment models (Section 7.3.1) and sensor data (Section 7.3.2), given a library of environment and correspondence models.

### 7.3.1   *Environment Model Encoding*

The environment model representation size, $\mathcal{L}[\mathbf{q}_E \,|\, \mathbf{Q}_E]$ (see Figure 7.3), depends on the model structure and number of parameters, and appropriate representation size formulae from Section 5.5 on page 66 are used, as dictated by the model representation.

Let $\mathbf{Q}_E(\Xi)$ denote the set of all possible instantiations of the model structure $\Xi$, i.e., $\mathbf{Q}_E(\Xi) \overset{\Delta}{=} \{\mathbf{q}_E : \mathbf{q}_E = \Xi(\boldsymbol{\theta}), \boldsymbol{\theta} \in \Theta\}$. An environment model library $\mathbf{Q}_E$ containing $L$ environment model structures may be partitioned into disjoint sets $\mathbf{Q}_E(\Xi)$:

$$\mathbf{Q}_E = \bigcup_{l=1}^{L} \mathbf{Q}_E(\Xi_l).$$

Often, the parameter set $\Theta$ is finite and can be discretized. The $l$-th structure with exactly $D_l$ independent parameters may be encoded in

$$\mathcal{L}[\mathbf{q}_E = \Xi^{(l)}(\boldsymbol{\theta}) \,|\, \mathbf{Q}_E] = \log_2(L+1) + \sum_{d=1}^{D_l} c_{\theta_d^{(l)}} \quad \text{bits} \qquad (7.11)$$

where the first term is the number of bits to specify the index of the model structure in the library, while the second term is the number of bits needed to encode the $D_l$ parameters. Assuming the $d$-th parameter lies in the finite interval, $\theta_d \in \Theta_d$, it can be discretized and encoded in:

$$c_{\theta_d^{(l)}} = \log_2\left(\frac{\text{Vol}(\Theta_d)}{\text{Vol}(\delta\theta_d)} + 1\right) \qquad (7.12)$$

bits, where $\delta\theta_d$ is a component of the parameter resolution (Equation 6.13). Alternatively, instead of a component-wise encoding of the environment model parameter vector, we could use a multi-dimensional counting scheme:

$$\mathcal{L}[\mathbf{q}_E = \Xi^{(l)}(\boldsymbol{\theta}) \,|\, \mathbf{Q}_E] = \log_2(L+1) + \log_2\left(\frac{\text{Vol}(\Theta)}{\text{Vol}(\delta\boldsymbol{\theta})} + 1\right) \qquad (7.13)$$

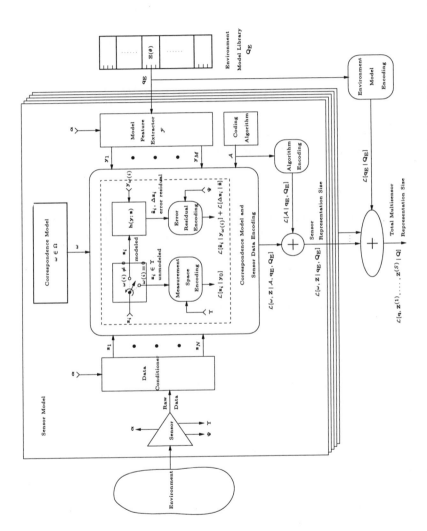

**Fig. 7.3** Minimal representation size multisensor fusion and model selection framework. The dotted line inset shows the encoding of a single data feature.

where $\delta\theta$ is a function of the multiple sensor accuracies, and sensor resolutions. In either encoding scheme, the model size depends inversely on the parameter resolution. Therefore the best model size typically increases as the sensor accuracy, $\Lambda$, decreases, or as the number of data features increase for the various sensors.

For unbounded model parameters integer coding schemes may be used (see Rissanen [166]):

$$\mathcal{L}[\mathbf{q}_{\mathrm{E}} = \Xi^{(l)}(\boldsymbol{\theta}) \,|\, \mathbf{Q}_{\mathrm{E}}] = \log_2(L+1) + \mathcal{I}(\frac{\mathrm{Vol}(\boldsymbol{\theta})}{\mathrm{Vol}(\delta\boldsymbol{\theta})}) \qquad (7.14)$$

where we used a spiral counting scheme (Section 6.3.1) with respect to some arbitrarily chosen origin in the parameter space.

### 7.3.2  *Sensor Encoding*

The sensor representation size, $\mathcal{L}[\omega, \mathbf{Z} \,|\, \mathbf{q}_{\mathrm{E}}, \mathbf{Q}_{\mathrm{E}}] \equiv \mathcal{L}[\omega, \mathbf{Z} \,|\, \mathbf{Y}]$ (see Figure 7.3), is developed next. We start by considering the representation size of a single data feature, and use it to build the representation size expressions which include the correspondence and the encoding algorithm.

#### 7.3.2.1  *Data Feature Representation Size*

The representation size of the observed data features depends on the subsampling of data to identify those which are regarded as "modeled", that is, having correspondence to the model features. In practice, this distinction may be interpreted as the identification of statistical "outliers", or noisy data features which are not likely to have resulted from the environment model.

The representation size $\mathcal{L}[\mathbf{z} \,|\, \mathbf{y}]$ of an observed data feature $\mathbf{z}$, when it is modeled ($\mathbf{y} \neq \mathbf{y}_0$), and when it is unmodeled ($\mathbf{y} = \mathbf{y}_0$), is developed next, and is summarized in Figure 7.3.

**Modeled Data Features**  The representation size of modeled data ($\omega(i) \neq 0$ in Figure 7.3) is based on an encoding of the error residuals between the data features and the corresponding model features, relative to the sensor constraint equation.

The *data constraint manifold* (Figure 7.4) associated with a model feature $\mathbf{y}$, is the set of "ideal" data features that can arise from

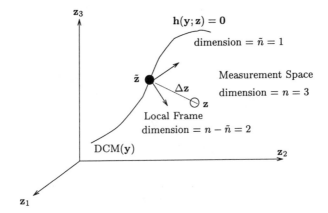

Fig. 7.4 A 1-dimensional DCM embedded in a 3-dimensional measurement space.

that model feature:

$$\text{DCM}(\mathbf{y}) \triangleq \{\mathbf{z} : \mathbf{h}(\mathbf{y};\mathbf{z}) = \mathbf{0}\}.$$

As shown in Figure 7.4, an observed data feature $\mathbf{z}$ can be encoded by specifying an ideal data feature $\tilde{\mathbf{z}}$ on the DCM, and the *error residual* $\Delta\mathbf{z} \triangleq \mathbf{z} - \tilde{\mathbf{z}}$:

$$\mathcal{L}[\mathbf{z}\,|\,\mathbf{y}] = \mathcal{L}[\tilde{\mathbf{z}}\,|\,\mathbf{y}] + \mathcal{L}[\Delta\mathbf{z}\,|\,\tilde{\mathbf{z}}]. \tag{7.15}$$

Let us assume the DCM defines an $\tilde{n}$-dimensional *parameterizable* surface embedded in an $n$-dimensional measurement space. We require exactly $\tilde{n}$ independent variables to "address" every point on the DCM; they can be chosen as some $\tilde{n}$ independent variables in the measurement space, as shown in Figure 7.5. Assuming that the $k$-th variable can be encoded in $\tilde{c}_k$ bits, the number of bits needed to address any point on the constraint manifold is given by

$$\mathcal{L}[\tilde{\mathbf{z}}\,|\,\mathbf{y}] = \sum_{k=1}^{\tilde{n}} \tilde{c}_k. \tag{7.16}$$

A point $\tilde{\mathbf{z}}$ on the DCM can be regenerated by specifying $\tilde{n}$ *independent* variables and the constraint equation $\mathbf{h}(\mathbf{y};\mathbf{z}) = \mathbf{0}$, as shown in Figure 7.5. When the DCM is a single point, i.e., $\tilde{n} = 0$, $\tilde{\mathbf{z}} = \mathbf{y}$, the DCM representation size is zero: $\mathcal{L}[\tilde{\mathbf{z}}\,|\,\mathbf{y}] = 0$.

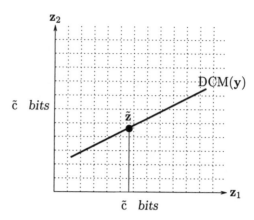

Fig. 7.5   Representation size of a 1-dimensional DCM embedded in a 2-dimensional measurement space, is the number of bits needed to specify a point $\tilde{z}$ on the DCM.

The error residual, $\Delta z$, can be regenerated by specifying the remaining $n - \tilde{n}$ *independent* variables, in a local reference frame *orthogonal* to the DCM at $\tilde{z}$, as shown in Figure 7.4. Given an uncertainty model describing the error residual in this *measurement subspace locally orthogonal to the DCM*, $\Delta z$ can be encoded as described in Section 6.3 (Equations 6.3 and 6.8).

In general, the ideal data feature $\tilde{z}$ is chosen to be the one that minimizes the representation size,

$$\tilde{z} \overset{\Delta}{=} \arg \min_{z \in \mathrm{DCM}(y)} \mathcal{L}[\Delta z \,|\, z]. \tag{7.17}$$

Often the closest point on the DCM is chosen. An algorithm for computing the error residual, $\Delta z$, is presented in Appendix C.

**Unmodeled Data Features**   Unmodeled, or "outlier" data features ($\omega(i) = 0$ in Figure 7.3) are unconstrained, and may lie anywhere in the measurement space. Given a description of the measurement space, $\Upsilon$, their representation size, is given by the number of bits needed to encode a point in this measurement space, $\mathcal{L}[z \,|\, y_0]$.

For measurement spaces described by the sensor accuracy and precision, the unmodeled data feature representation size is given by the *sensor precision*

$$\mathcal{L}[z \,|\, y_0] = \lambda$$

as discussed in Section 6.3.1. For a measurement space described by a probability distribution, the unmodeled data feature representation size is given by

$$\mathcal{L}[\mathbf{z} \mid \mathbf{y}_0] = -\log_2 p_0(\mathbf{z}) - \sum_{k=1}^n \log_2 \delta_k$$

or by the average sensor precision, $\bar{\lambda}$ (Equation 6.10), as discussed in Section 6.3.2.

The representation size of an observed data feature is bounded above by its outlier representation size; thus

$$\mathcal{L}[\mathbf{z} \mid \mathbf{y}] \le \mathcal{L}[\mathbf{z} \mid \mathbf{y}_0], \qquad (7.18)$$

since $\mathbf{z}$ can always be treated as an outlier by choosing $\mathbf{y} = \mathbf{y}_0$ (unless the correspondence is already known and $\mathbf{y}$ cannot be chosen). Therefore, the unmodeled representation size, $\mathcal{L}[\mathbf{z} \mid \mathbf{y}_0]$, controls the balance between the modeled and the unmodeled data features. For ellipsoidal uncertainty regions, it may be viewed as specifying the radius of an error ellipsoid around around each point on the D-CM surface; only the data features inside these hyperspheres are candidates for being modeled.

### 7.3.2.2 *Two-Part Encoding*

In *two-part* encoding the data and correspondence are encoded separately (Section 7.2.1.1):

$$\mathcal{L}[\omega, \mathbf{Z} \mid \mathcal{A}, \mathbf{q}_{\mathrm{E}}, \mathbf{Q}_{\mathrm{E}}] = \mathcal{L}[\omega \mid \mathcal{A}, \mathbf{q}_{\mathrm{E}}, \mathbf{Q}_{\mathrm{E}}] + \sum_{i=1}^N \mathcal{L}[\mathbf{z}_i \mid \mathbf{y}_{\omega(i)}],$$

where the individual data feature representation size, $\mathcal{L}[\mathbf{z}_i \mid \mathbf{y}_{\omega(i)}]$, is calculated as described in Section 7.3.2.1. The representation size expressions for encoding the correspondence, $\mathcal{L}[\omega \mid \mathcal{A}, \mathbf{q}_{\mathrm{E}}, \mathbf{Q}_{\mathrm{E}}]$, are developed next.

A correspondence, $\omega$, specifies for each data feature, the index of the corresponding model feature. It may be viewed as a $N$ *letter word from a* $(M + 1)$-*ary alphabet* ($M$ model features, plus the symbol 0 to denote outliers). Let $N_j$, $j = 0, \dots, M$ denote the number of occurrences of the value $j$ in $\omega$. Let us define the *relative frequency* of occurrence of model feature $\mathbf{y}_j$ as, $f_j \triangleq N_j/N$, $j = 0, \dots, M$ and denote $\mathbf{f} = [f_0 \dots, f_M]^{\mathrm{T}}$.

The model feature id may be regarded as a discrete $(M+1)$-ary random variable, and the relative frequency vector as a *probability mass function* associated with the $(M+1)$ symbols $\mathbf{y}_0, \ldots, \mathbf{y}_M$. For a discrete $K$-ary random variable, its *entropy* [40],

$$H(p_1, \ldots, p_K) \overset{\Delta}{=} -\sum_{k=1}^{K} p_k \log_2 p_k$$

gives the shortest average codelength per symbol. Since there are exactly $N_j$ occurrences of $\mathbf{y}_j$ in $\omega$, the shortest (optimal) codelength of $\omega$ is

$$\mathcal{L}[\omega \,|\, \mathcal{A}, \mathbf{q}_{\mathrm{E}}, \mathbf{Q}_{\mathrm{E}}] = N \cdot H(f_0, \ldots, f_M). \tag{7.19}$$

Equation (7.19) can be also derived by counting the number of ways $N$ objects can be assigned $M+1$ labels. Given that $N_j$ objects are assigned the $j$th label, the total number of ways is:

$$\frac{N!}{N_0! \ldots N_M!} \approx \frac{N^N}{N_0^{N_0} \ldots N_M^{N_M}}$$

where we used the *Stirlings approximation*, $N! \approx (N/e)^N \sqrt{2\pi e N}$. A correspondence assignment is fully specified by its index in a complete (lexicographic) ordering of the set of all legal correspondences, $\Omega$. Therefore, the number of bits needed to completely specify $\omega \in \Omega$ is given by (Equation 5.8 on page 67):

$$\log_2 \frac{N!}{N_0! \ldots N_M!} \approx N \cdot H(f_0, \ldots, f_M)$$

since $N_j = N \cdot f_j$. This is same as Equation (7.19).

In this encoding scheme, the relative frequency $\mathbf{f}$ is a variable, and must be supplied as an argument to the encoding algorithm; it comprises the algorithm representation size, $\mathcal{L}[\mathcal{A} \,|\, \mathbf{q}_{\mathrm{E}}, \mathbf{Q}_{\mathrm{E}}]$. Since $f_j \in [0,1]$, and the smallest increment in $f_j$ is in steps of $1/N$, therefore the number of bits needed to describe all possible values of $f_j$ is $\log_2(\frac{1}{1/N}+1) \approx \log_2 N$, using Equation (5.19). Since $\sum_{j=0}^{M} f_j = 1$, at most $M$ *variable* elements of the $\mathbf{f}$ need to be explicitly specified; therefore

$$\mathcal{L}[\mathcal{A} \,|\, \mathbf{q}_{\mathrm{E}}, \mathbf{Q}_{\mathrm{E}}] = c_{\mathcal{A}} + \sum_{f_j \, \mathrm{variable}} \log_2 N \tag{7.20}$$

where $c_{\mathcal{A}}$ is the constant number of bits needed to specify the encoding algorithm "preamble" (includes the algorithm text and the value of $N$), and the second term is the number of bits needed to specify its variable arguments.

**Many-to-One Correspondence** In a many-to-one correspondence (Section 6.2.1), all combinations on the $(M + 1)$ symbols in the $N$-letter correspondence word are legal. Therefore, simplifying Equation (7.19), we get

$$\mathcal{L}[\omega \,|\, \mathcal{A}, \mathbf{q}_{\mathrm{E}}, \mathbf{Q}_{\mathrm{E}}] = -\sum_{j=0}^{M} N_j \log_2 f_j \equiv -\sum_{i=1}^{N} \log_2 f_{\omega(i)}.$$

The last term is just a rearrangement of indices, and can be rewritten as

$$\mathcal{L}[\omega \,|\, \mathcal{A}, \mathbf{q}_{\mathrm{E}}, \mathbf{Q}_{\mathrm{E}}] = N \log_2 N - \sum_{i=1}^{N} \log_2 N_{\omega(i)}. \qquad (7.21)$$

We can interpret this expression as:

*Exactly $\log_2 N - \log_2 N_{\omega(i)}$ bits are needed for the $i$-th data feature, to encode a many-to-one correspondence.*

In this case, exactly $M$ elements of the relative frequency, $\mathbf{f}$, are variable; therefore, from Equation (7.20), we have:

$$\mathcal{L}[\mathcal{A} \,|\, \mathbf{q}_{\mathrm{E}}, \mathbf{Q}_{\mathrm{E}}] = c_{\mathcal{A}} + M \cdot \log_2 N \qquad (7.22)$$

as the algorithm representation size.

**One-to-One Correspondence** In a one-to-one correspondence (Section 6.2.2), not all combinations on the $(M + 1)$ symbols in the $N$-letter correspondence word are legal. There can be *at most* one symbol $\mathbf{y}_j, j = 1, \ldots, M$ in $\omega$, although $\mathbf{y}_0$ may be repeated several times. Let $\tilde{N} = N - N_0$ denote the number of *modeled* data features, $\tilde{N} \leq \min(M, N)$. Therefore,

$$f_0 = 1 - \frac{\tilde{N}}{N} \in [1 - \frac{\min(M, N)}{N}, 1],$$

due to the $N_0$ *outlier* data features, and

$$f_j = \frac{1}{N} = \frac{(1 - f_0)}{\tilde{N}},$$

for exactly $\tilde{N}$ model features corresponding to the $\tilde{N}$ *modeled* data features, and

$$f_j = 0,$$

for the remaining $M - \tilde{N}$ *missing* model features.

Putting these constraints on $\mathbf{f}$, in Equation (7.19), we get the representation size for a one-to-one correspondence:

$$
\begin{aligned}
\mathcal{L}[\omega \,|\, \mathcal{A}, \mathbf{q}_{\mathrm{E}}, \mathbf{Q}_{\mathrm{E}}] &= N \cdot H(f_0, \overbrace{1/N, \ldots, 1/N}^{\tilde{N}}, \overbrace{0, \ldots, 0}^{M-\tilde{N}}) \\
&= -N_0 \log_2 f_0 - \tilde{N} \log_2(1/N) \\
&= N \log_2 N - N_0 \log_2 N_0.
\end{aligned}
$$

This can be rewritten as:

$$\mathcal{L}[\omega \,|\, \mathcal{A}, \mathbf{q}_{\mathrm{E}}, \mathbf{Q}_{\mathrm{E}}] = \tilde{N} \log_2 N + N_0 \log_2 N - N_0 \log_2 N_0, \qquad (7.23)$$

which can be interpreted as:

> *Exactly* $\log_2 N - \log_2 N_0$ *bits are needed per unmodeled data feature, while* $\log_2 N$ *bits are needed per modeled data feature, to encode a one-to-one correspondence.*

In this case, just one parameter, namely $f_0$, is variable; therefore, from Equation (7.20), we have:

$$\mathcal{L}[\mathcal{A} \,|\, \mathbf{q}_{\mathrm{E}}, \mathbf{Q}_{\mathrm{E}}] = c_{\mathcal{A}} + \log_2 N. \qquad (7.24)$$

**Practical Approximations**  A *fixed-length* encoding scheme specifies, for each data feature, the id of the corresponding model feature, using exactly $\log_2(M + 1)$ bits:

$$\mathcal{L}[\omega \,|\, \mathcal{A}, \mathbf{q}_{\mathrm{E}}, \mathbf{Q}_{\mathrm{E}}] = N \log_2(M + 1). \qquad (7.25)$$

In this case, the encoding algorithm does not require any arguments, and is specified by a fixed number of bits:

$$\mathcal{L}[\mathcal{A} \,|\, \mathbf{q}_{\mathrm{E}}, \mathbf{Q}_{\mathrm{E}}] = c_{\mathcal{A}}. \qquad (7.26)$$

This fixed-length encoding can be viewed as a special case of Equation (7.19), when all the $f_j$ are equal, $f_j = 1/(M+1)$, $j = 0 \ldots M$. The entropy is maximum for this distribution [40], and therefore this fixed-length encoding gives an *upper bound* for the Equations (7.21) and (7.23).

The fixed-length encoding is preferred for small $N$, i.e., when $N \leq M$, since its correspondence plus algorithm codelength, $c_A + N \cdot \log_2(M + 1)$, is smaller than the codelength, $c_A + M \cdot \log_2 N + N \cdot H(f_0, \ldots, f_M)$, for the entropy based encoding scheme. As $N$ increases $(N >> M)$, the entropy based encoding gives smaller codelengths, and is preferred. In practice, we use that encoding scheme which gives the smaller codelengths, and specify the identity of the actual coding algorithm in the preamble (the $c_A$ bits).

For two-part encoding, this fixed-length encoding is especially convenient, since it simplifies the representation size calculation, and therefore the correspondence search algorithm (Section 8.3 on page 128).

### 7.3.2.3 *Mixture Encoding*

In *mixture* encoding, the correspondence and data are encoded *jointly* (Section 7.2.1.1). The mixture representation size expressions, $\mathcal{L}[\omega, \mathbf{z} \mid \mathcal{A}, \mathbf{q}_E, \mathbf{Q}_E]$, for the joint encoding of the correspondence $\omega$ and a data feature $\mathbf{z}$, are developed next.

The data feature representation size (Section 7.3.2.1) can be rewritten as $\mathcal{L}[\mathbf{z} \mid \mathbf{y}] = -\log_2 \mathrm{P}(\mathbf{z} \mid \mathbf{y})$, where,

$$\mathrm{P}(\mathbf{z} \mid \mathbf{y}) \triangleq 2^{-\log_2 \mathcal{L}[\mathbf{z} \mid \mathbf{y}]} \tag{7.27}$$

is defined to be a *natural* or *universal* distribution on the observed data features, induced by the representation size. Note that this distribution must satisfy the Kraft inequality (Equation 5.1 on page 59), since prefix codes are used; therefore it can always be normalized to form a proper probability distribution of area unity.

An observed data feature $\mathbf{z}_i$ may correspond to one of the model features $\mathbf{y}_0, \ldots, \mathbf{y}_M$, and may be regarded as a "mixture" of $M + 1$ random variables, random variables mixed according to another *independent* discrete random variable $\omega(i)$, whose probability mass function is $\mathbf{f}$. The mixture representation size is given by the number of bits required to encode this mixture distribution (Appendix D):

$$\mathcal{L}[\omega, \mathbf{z} \mid \mathcal{A}, \mathbf{q}_{\mathrm{E}}, \mathbf{Q}_{\mathrm{E}}] = -\log_2 \left( \sum_{j=0}^{M} f_j \cdot 2^{-\mathcal{L}[\mathbf{z} \mid \mathbf{y}_j]} \right) \tag{7.28}$$

where $\mathcal{L}[\mathbf{z} \mid \mathbf{y}_j]$ are obtained as discussed in Section 7.3.2.1 on page 100.

Using a mixture encoding results in shorter codelengths than two-part encoding:

$$\mathcal{L}[\omega \mid \mathcal{A}, \mathbf{q}_{\mathrm{E}}, \mathbf{Q}_{\mathrm{E}}] + \mathcal{L}[\mathbf{Z} \mid \omega, \mathcal{A}, \mathbf{q}_{\mathrm{E}}, \mathbf{Q}_{\mathrm{E}}]$$

$$= -N \cdot \sum_{j=0}^{M} f_j \cdot \log_2 f_j + \sum_{i=1}^{N} \mathcal{L}[\mathbf{z}_i \mid \mathbf{y}_{\omega(i)}]$$

$$= -\sum_{i=1}^{N} \log_2 f_{\omega(i)} - \sum_{i=1}^{N} \log_2 \mathrm{P}(\mathbf{z}_i \mid \mathbf{y}_{\omega(i)})$$

$$= -\sum_{i=1}^{N} \log_2 f_{\omega(i)} \cdot \mathrm{P}(\mathbf{z}_i \mid \mathbf{y}_{\omega(i)})$$

$$\geq -\sum_{i=1}^{N} \log_2 \sum_{j=0}^{M} f_j \cdot \mathrm{P}(\mathbf{z}_i \mid \mathbf{y}_j)$$

$$= \mathcal{L}[\omega, \mathbf{Z} \mid \mathcal{A}, \mathbf{q}_{\mathrm{E}}, \mathbf{Q}_{\mathrm{E}}]$$

where the inequality follows from the properties of the $\log_2(\cdot)$ function (Appendix D). The savings result from the fact that, rather than assigning a separate codeword per component distribution, $\mathrm{P}(\mathbf{z} \mid \mathbf{y})$, as is the case with two-part encoding, a single codeword is shared for those values of $\mathbf{z}$ for which the component distributions overlap. A single codeword simultaneously identifies the correspondence $\omega(i)$ and $\Delta\mathbf{z}_i$, which may be used to regenerate $\mathbf{z}_i$. However, the correspondence is no longer explicitly represented, but implicit in the relative frequency vector $\mathbf{f}$.

In the mixture encoding scheme, as with two-part encoding, the relative frequency $\mathbf{f}$ is a variable, and must be supplied as an argument to the encoding algorithm. Therefore, the algorithm representation size, $\mathcal{L}[\mathcal{A} \mid \mathbf{q}_{\mathrm{E}}, \mathbf{Q}_{\mathrm{E}}]$ is given by (Equation 7.20):

$$\mathcal{L}[\mathcal{A} \mid \mathbf{q}_{\mathrm{E}}, \mathbf{Q}_{\mathrm{E}}] = c_{\mathcal{A}} + \sum_{f_j \text{variable}} \log_2 N. \tag{7.29}$$

In Appendix E, we show that the mixture representation size is strictly convex $\cup$ in $\mathbf{f}$, and lies in the range:

$$-\log_2(f_0 \cdot 2^{-\mathcal{L}[\mathbf{z}\,|\,\mathbf{y}_0]} + 1 - f_0)) \leq \mathcal{L}[\omega, \mathbf{z}\,|\,\mathbf{Y}] \leq -\log_2 f_0 + \mathcal{L}[\mathbf{z}\,|\,\mathbf{y}_0]. \tag{7.30}$$

Note that as $f_0 \to 0$, the bounds stretch to the range $[0, \infty)$; and as $f_0 \to 1$, the bounds contract and approach $\mathcal{L}[\mathbf{z}\,|\,\mathbf{y}_0]$, the unmodeled data feature representation size.

**Many-to-One Correspondence** In a many-to-one correspondence (Section 6.2.1), all combinations on the $(M+1)$ symbols in the $N$-letter correspondence word are legal. Therefore, from Equation (7.28), we have the mixture representation size

$$\mathcal{L}[\omega, \mathbf{z}\,|\,\mathcal{A}, \mathbf{q}_{\mathrm{E}}, \mathbf{Q}_{\mathrm{E}}] = -\log_2\left(\sum_{j=0}^{M} f_j \cdot 2^{-\mathcal{L}[\mathbf{z}\,|\,\mathbf{y}_j]}\right) \tag{7.31}$$

and from Equation (7.29), we have the algorithm representation size

$$\mathcal{L}[\mathcal{A}\,|\,\mathbf{q}_{\mathrm{E}}, \mathbf{Q}_{\mathrm{E}}] = c_{\mathcal{A}} + M \cdot \log_2 N. \tag{7.32}$$

**One-to-One Correspondence** In a one-to-one correspondence, not all combinations on the $(M + 1)$ symbols in the $N$-letter correspondence word are legal, and the resulting constraints on the relative frequency vector, $\mathbf{f}$, are described in Section 7.3.2.2. Substituting in Equation (7.28), the mixture representation size is given by

$$\mathcal{L}[\omega, \mathbf{z}\,|\,\mathcal{A}, \mathbf{q}_{\mathrm{E}}, \mathbf{Q}_{\mathrm{E}}]$$
$$= -\log_2\left(f_0 \cdot 2^{-\mathcal{L}[\mathbf{z}\,|\,\mathbf{y}_0]} + \frac{1}{N} \cdot \sum_{j=1}^{M} \delta(N_j - 1) \cdot 2^{-\mathcal{L}[\mathbf{z}\,|\,\mathbf{y}_j]}\right) \tag{7.33}$$

where $\delta(\cdot)$ is the Dirac-Delta function. The algorithm representation size is given by

$$\mathcal{L}[\mathcal{A}\,|\,\mathbf{q}_{\mathrm{E}}, \mathbf{Q}_{\mathrm{E}}] = c_{\mathcal{A}} + \log_2 N. \tag{7.34}$$

since only $f_0$ is variable.

**Practical Approximations** Using the *fixed-length* encoding scheme, described in Section 7.3.2.2, the mixture representation size is given by

$$\mathcal{L}[\omega, \mathbf{z} \mid \mathcal{A}, \mathbf{q}_{\mathrm{E}}, \mathbf{Q}_{\mathrm{E}}] = -\log_2 \left( \sum_{j=0}^{M} \frac{2^{-\mathcal{L}[\mathbf{z} \mid \mathbf{y}_j]}}{M+1} \right)$$

As for the two-part encoding case, the encoding algorithm does not require any arguments, and is specified by a fixed number of bits:

$$\mathcal{L}[\mathcal{A} \mid \mathbf{q}_{\mathrm{E}}, \mathbf{Q}_{\mathrm{E}}] = c_{\mathcal{A}}.$$

A more useful approximation for mixture encoding*, is obtained by putting $f_j = (1 - f_0)/M$, $j = 1 \ldots M$ for all the *modeled* data features, so that only the *fraction of outliers*, $f_0$, is variable. The mixture representation size is given by

$$\mathcal{L}[\omega, \mathbf{z} \mid \mathcal{A}, \mathbf{q}_{\mathrm{E}}, \mathbf{Q}_{\mathrm{E}}] = -\log_2 \left( f_0 \cdot 2^{-\mathcal{L}[\mathbf{z} \mid \mathbf{y}_0]} + \frac{1 - f_0}{M} \cdot \sum_{j=1}^{M} 2^{-\mathcal{L}[\mathbf{z} \mid \mathbf{y}_j]} \right)$$

$$(7.35)$$

while the algorithm representation size is given by

$$\mathcal{L}[\mathcal{A} \mid \mathbf{q}_{\mathrm{E}}, \mathbf{Q}_{\mathrm{E}}] = c_{\mathcal{A}} + \log_2 N. \tag{7.36}$$

This approximation is especially convenient in practice, since it allows estimating the percentage of outliers in the data, while keeping the search space small (only $f_0$ must be estimated, instead of the complete relative frequency vector $\mathbf{f}$).

### 7.3.3 *Choosing an Encoding Scheme*

The minimal representation size multisensor fusion and model selection framework (Figure 7.3) allows us to fuse disparate sensor data using the representation size as a common sensor independent universal yardstick, which measures the "information" inherent in observed data relative to a class of environment and sensor models. The various sensor data and correspondence encoding schemes, developed in Section 7.3.2, are summarized

---

*For two-part encoding, this approximation gives a tighter upper bound for Equations (7.21) and (7.23), than the fixed-length encoding.

Table 7.1  Sensor encoding schemes.

| $\mathcal{L}[\omega, \mathbf{Z} \mid \mathbf{q}_E, \mathbf{Q}_E]$ | $\mathcal{L}[\mathcal{A} \mid \mathbf{q}_E, \mathbf{Q}_E]$ | + | $\mathcal{L}[\omega, \mathbf{Z} \mid \mathcal{A}, \mathbf{q}_E, \mathbf{Q}_E]$ |
|---|---|---|---|
| Mixture Encoding | $c_{\mathcal{A}} + \sum_{f_j \text{ variable}} \log_2 N$ | + | $-\sum_{i=1}^{N} \log_2\{\sum_{j=0}^{M} f_j \cdot 2^{-\mathcal{L}[\mathbf{z}_i \mid \mathbf{y}_j]}\}$ |
| Two-Part Encoding | $c_{\mathcal{A}} + \sum_{f_j \text{ variable}} \log_2 N$ | + | $\sum_{i=1}^{N}\{-\log_2 f_{\omega(i)} + \mathcal{L}[\mathbf{z}_i \mid \mathbf{y}_{\omega(i)}]\}$ |
| Known Correspondence | $0$ | + | $\sum_{i=1}^{N} \mathcal{L}[\mathbf{z}_i \mid \mathbf{y}_{\omega(i)}]$ |

Table 7.2  Data feature representation size.

| $\mathcal{L}[\mathbf{z} \mid \mathbf{y}]$ | $\mathcal{L}[\bar{\mathbf{z}} \mid \mathbf{y}]$ + | Modeled, $\mathbf{y} \neq \mathbf{y}_0$ $\mathcal{L}[\Delta\mathbf{z} \mid \bar{\mathbf{z}}]$ | Unmodeled, $\mathbf{y} = \mathbf{y}_0$ $\mathcal{L}[\mathbf{z} \mid \mathbf{y}_0]$ |
|---|---|---|---|
| Probabilistic | $\sum_{k=1}^{\bar{n}} \bar{c}_k$ + | $-\log_2 \mathrm{p}(\Delta\mathbf{z} \mid \bar{\mathbf{z}}) - \sum_{k=1}^{n} \log_2 \delta_k$ | $-\log_2 \mathrm{p}_0(\mathbf{z}) - \sum_{k=1}^{n} \log_2 \delta_k$ |
| Accuracy and Precision | $\sum_{k=1}^{\bar{n}} \bar{c}_k$ + | $\log_2\left(\frac{\mathrm{Vol}(\Delta\mathbf{z})}{\mathrm{Vol}(\Lambda)} + 1\right)$ | $\log_2\left(\frac{\mathrm{Vol}(\Upsilon)}{\mathrm{Vol}(\Lambda)} + 1\right)$ |

in Table 7.1. Table 7.2 summarizes the data feature representation size expressions for the different uncertainty model representations. For each sensor, the appropriate encoding alternatives are chosen, and used in the expression for the total multisensor representation size (Equation 7.10). Some considerations for choosing between these alternatives are discussed below.

The choice between **two-part vs. mixture encoding** for a sensor, is dictated by the need for estimating the discrete correspondence, and the typical size of an observed data set. Let us denote the set of discrete correspondences that result in the same relative frequency distribution, **f**, by $\Omega(\mathbf{f})$. The size of $\Omega(\mathbf{f})$ is given by:

$$| \Omega(\mathbf{f}) | = \frac{N!}{N_0! \ldots N_M!} \approx e^{N \cdot H(f_0, \ldots, f_M)}$$

where $N_j = N \cdot f_j$, and we used the Stirlings approximation (page 104). Thus, as $N$ increases, the size of $\Omega(\mathbf{f})$ *grows exponentially*, while the quantization of $\mathbf{f}$ decreases as $1/N$. Therefore, from a computational standpoint, mixture encoding should be used for large data sets, since the dimensionality of its search space is independent of the number of data features and it is convex in $\mathbf{f}$ (Appendix E). However, we can no longer distinguish between the different elements of $\omega \in \Omega(\mathbf{f})$, as we could with two-part encoding.

The choice between **many-to-one vs. one-to-one correspondence** for a sensor, is usually dictated by the characteristics of the specific sensor under consideration. When there is no clear dictate, the many-to-one correspondence is preferable due to fewer constraints on the set legal correspondences. This usually simplifies the search procedures (Section 8.3).

The choice between **probabilistic vs. encoded accuracy and precision** uncertainty models for a sensor, is usually dictated by its observation error characteristics, and the ability to build reliable probabilistic error models. In cases when a probability distribution is difficult to define or approximate, it is possible to use the accuracy of the sensors and the precision of the data as a basis for definition of the encoded representation size. For some sensors (such as the Vision sensor), the accuracy and precision constitute a more "natural" representation of sensor uncertainty. Probabilistic uncertainty models can also be mapped into sensor accuracy and precision, as discussed in Section 6.3.2.1 on page 87.

It should be obvious that additional *non-numeric symbolic attributes*, can be easily incorporated into the multisensor representation size of Equation (7.10), by regarding them as a new type of sensor and including additional additive terms for their representation size (computed using the encoding schemes discussed in Section 5.5).

## 7.4   Model Selection Properties

### 7.4.1   *Environment Model Class Selection*

Historically, the minimal representation size techniques were developed to address model selection issues, where the model class and the number of model parameters must be selected. A typical example of selecting the number of model parameters is to find the degree of the polynomial that best describes the observed data (Section 1.3.2). The related issue of model class selection address the choice between, say, the family of polynomials vs. transcendental functions. Other researchers have addressed these topics, and have amply demonstrated the use of the minimal representation size as a criterion for selecting the model class and the number of parameters [1, 13, 168–170, 177].

For pure model class selection problems (such as the object recognition problems discussed in Part 3), the number of model parameters is the same for all environment model structures in the model library, and only the model class and the parameter resolution must be selected. In these problems, the correspondence representation size (Table 7.1 on page 111) contributes a term which favors models with smaller number of model features, $M$. This term also depends on the number of data features, $N$, so that sensors with with more data features have a greater influence in choosing the model class. The simplest model in the sense of having the smallest "overall" number of model features ("weighted" according to the number of data features), that best describes the observed multisensor data is chosen from a library of environment models. This property is illustrated by the experimental results presented in Section 13.1.3 (Table 13.1).

### 7.4.1.1 *Environment Model Parameter Resolution*

The "best" model is automatically selected by the minimal representation size criterion for a given set of observed multisensor data (Equation 7.10). In general, this includes the selection of the model class (or id), the number of model parameters, and the number of bits per model parameter. Since, the model size expressions (Equations 7.11, 7.12, and 7.13), depend inversely on the parameter resolution of the sensors, the selected model (parameter) size implicitly defines an "optimal" parameter resolution for an observed data set. In practice, this "optimal" parameter resolution may be found by searching through the truncations of the parameter values, obtained from an estimation algorithm [177].

Alternatively, we could derive an expression for the parameter resolution, as a function of the sensor accuracies and the number of data features, and use it to define the "best" model size. Using the Cramer-Rao inequality [40, page 328], a lower bound on the parameter resolution, $\delta\boldsymbol{\theta}$, is given by the inverse *Fisher information*. This can be used as basis for the definition of the "optimal" parameter resolution, which results in the "best" model size according to Equations (6.13) and (7.13). As the Fisher information increases, this "best" model size increases, and results in improved model parameter resolution.

Therefore, in this framework, the resulting model parameter resolution is controlled by the **sensor resolution** and the **sensor accuracy** of the various sensors—higher sensor resolutions and sensor accuracies lead to improved model parameter resolution (larger model size).

### 7.4.2    *Environment Model Parameterization and Data Scaling*

The modeled data features, define the environment model parameters via the sensor constraint equations (Equation 6.1 on page 76), given the model class and number of parameters.

To characterize the model parameterization and data scaling in the minimal representation size multisensor fusion framework, let us consider the derivative of the representation size of a sensor w.r.t. the environment model parameters $\theta$, assuming a known correspondence and two-part encoding:

$$\frac{d\mathcal{L}[\omega, \mathbf{Z} \mid \mathbf{q}_E, \mathbf{Q}_E]}{d\theta} = \sum_{\omega(i) \neq 0} \frac{d\mathcal{L}[\mathbf{z}_i \mid \tilde{\mathbf{z}}_i]}{d\theta},$$

which depends only on the modeled data features, i.e., $\omega(i) \neq 0$. Its magnitude determines the influence on model parameter selection—the larger the magnitude, the greater the local influence of these sensor data features on the model parameterization, and vice-versa. Its sign determines the direction of minimization.

We can write each term in the above as,

$$\frac{d\mathcal{L}[\mathbf{z} \mid \tilde{\mathbf{z}}]}{d\theta} = \frac{d\mathcal{L}[\mathbf{z} \mid \tilde{\mathbf{z}}]}{d\mathbf{z}} \cdot \frac{d\mathbf{z}}{d\theta} = \rho(\theta) \cdot \frac{d\mathcal{L}[\mathbf{z} \mid \tilde{\mathbf{z}}]}{d\mathbf{z}},$$

where $\rho(\theta)$ is the sensor resolution defined in Section 6.4 on page 89. Assuming the sensor resolution is independent of the data feature values $\mathbf{z}$, the total multisensor representation size is given by (Equation 7.10 on page 97):

$$\frac{d\mathcal{L}[\mathbf{q}, \mathbf{Z}^{(1)}, \dots, \mathbf{Z}^{(S)} \mid \mathbf{Q}]}{d\theta} = \sum_{s=1}^{S} \left\{ \rho_s(\boldsymbol{\theta}) \cdot \sum_{\omega(i) \neq 0} \frac{d\mathcal{L}[\mathbf{z}_i \mid \tilde{\mathbf{z}}_i]}{d\mathbf{z}_i} \right\}. \qquad (7.37)$$

In other words, the influence of each sensor is weighted by its sensor resolution. Therefore, **sensor resolution** controls the balance between the different sensors in determining the *model parameterization*. The higher the sensor resolution, the greater its influence on the model parameter selection and vice-versa.

The $\frac{d\mathcal{L}[\mathbf{z}_i \mid \tilde{\mathbf{z}}_i]}{d\mathbf{z}_i}$ terms in Equation (7.37) depend on the error residuals, $\Delta \mathbf{z}$, and the sensor uncertainty model, given by the sensor accuracy (Equation 7.15 on page 101 and Equation 6.3 on page 83). The **sensor accuracy** controls the *data scaling* among different sensors, while the error residuals control the influence among various data features of a single sensor.

Error residuals of subsampled data features from multiple sensors, "scaled" on the basis of sensor accuracy, and "weighted" on the basis of sensor resolution, are used to define the environment model parameters according to the sensor constraint equations, in the minimal representation size multisensor fusion framework.

Intuitively, more accurate sensors have greater influence on the model parameterization and vice-versa. If the uncertainty regions for a sensor are small (or 0 in the limit), the representation size terms for its data features having non-zero error residuals, are large (or $\infty$ in the limit). Therefore, those data features would have the greatest influence in choosing the model parameters and the correspondences (in the limit, these data features must be matched exactly with zero error residuals).

Given two similar sensors with identical error residuals but different accuracies, the more accurate sensor (smaller uncertainty region) would contribute larger representation size terms; and therefore would bias the final choice of model parameters towards smaller error residuals for its data features.

### 7.4.2.1 *Error Residuals*

Let us illustrate the influence of the error residuals among the various data features of a single sensor. For a two-dimensional ellipsoidal uncertainty region,

$$\frac{d\mathcal{L}[\mathbf{z} \mid \tilde{\mathbf{z}}]}{d\mathbf{z}} \approx \frac{2A\Delta\mathbf{z}}{(\Delta\mathbf{z}^{\mathrm{T}} A\Delta\mathbf{z}) + 1},$$

where we used Equation (5.12). When $\Delta\mathbf{z} \to 0$, $\frac{d\mathcal{L}[\mathbf{z}_i \mid \tilde{\mathbf{z}}_i]}{d\mathbf{z}_i} \propto \Delta\mathbf{z}$; and for such data features, the sensor constraints are "weighed" proportional to $\Delta\mathbf{z}$ in defining the model parameters, thus producing an *averaging* effect. When $\Delta\mathbf{z} \to \infty$, $\frac{d\mathcal{L}[\mathbf{z}_i \mid \tilde{\mathbf{z}}_i]}{d\mathbf{z}_i} \propto \frac{1}{\Delta\mathbf{z}}$; and for such data features, the sensor constraints have negligible effect in defining the model parameters. Thus, we get an averaging effect on the model parameter estimate for small $\Delta\mathbf{z}$, and negligible effect for large $\Delta\mathbf{z}$ (which may be outliers). Eventually the data features with large $\Delta\mathbf{z}$ would get eliminated from the subsampled data set, and be treated as outliers.

In contrast, the commonly used *least mean square error* (LSE) approach to model parameter estimation minimizes a penalty function of the form, $\mathcal{L}[\mathbf{z} \mid \tilde{\mathbf{z}}] = \Delta\mathbf{z}^{\mathrm{T}} A\Delta\mathbf{z}$, which can be obtained by using a *Gaussian* probability distribution for the error uncertainty model (Section 6.3.2). For this parameter estimation method, $\frac{d\mathcal{L}[\mathbf{z}_i \mid \tilde{\mathbf{z}}_i]}{d\mathbf{z}_i} \propto \Delta\mathbf{z}$, and produces an *averaging* effect for all values of $\Delta\mathbf{z}$. Thus, we get an averaging effect no matter how large $\Delta\mathbf{z}$ is, and therefore the outliers (with large $\Delta\mathbf{z}$) may adversely affect the model parameter estimate. The adverse effect due to the outliers gets more severe, until they get eliminated from the subsampled data set (in which the outliers have been rejected).

Given a *known* correspondence model, the encoded ellipsoidal accuracy and precision based sensor uncertainty model tends to select model parameters which favor a few small error residuals over several medium error residuals, while an LSE approach has the opposite behavior. For example, given two data features with prespecified correspondences, an ellipsoidal accuracy and precision encoding scheme would select model parameters so that at least one of the errors residuals is "small", while the other one may be "large". On the other hand, a Gaussian pdf/LSE encoding scheme would select model parameters which minimize the average errors, and both the errors residuals may be "medium". An ellipsoidal accuracy and precision based encoding scheme selects model parameters which minimize *some* of the error residuals, and prefers partial "perfect" fits. A Gaussian pdf/LSE encoding scheme selects parameters which minimize the average error residuals, and prefers "average" fits.

More importantly, either encoding method (ellipsoidal accuracy and precision, or Gaussian pdf/LSE) may be used for model parameter estimation, and the minimal representation size framework can automatically select the correspondence models and the environment model at an appropriate parameter resolution (dependent on the observed data, the sensor uncertainties, and the sensor constraints).

### 7.4.3   *Data Subsample Selection*

The unmodeled data feature representation size can be viewed as "model size" term in this framework, since its encodes a data feature in the measurement space of the sensor. As discussed in Section 7.3.2.1 (Equation 7.18), the minimal representation size criterion trades-off between representing a data feature as unmodeled, or associating it with an available model feature and encoding only the error residuals. The set of available model features are constrained by the correspondence model (Section 6.2).

The tradeoff between the model size (number of bits needed to represent the unmodeled data features) and encoded error residuals (number of bits needed to encode the errors with respect to the model features) for each sensor results in the selection of a correspondence model, which specifies the subset of modeled data features, consistent with the environment model. This property is illustrated by the experimental results presented in Section 13.1.2 (Figure 13.1).

The unmodeled data feature representation size, which depends on the **sensor precision**, controls the effectiveness of data subsampling.

### 7.5   Theoretical Analysis

### 7.5.1   *Information Gain*

We define the notion of "information gain" to quantify the amount of information obtained from additional data, and use it to show that using additional sensors does not diminish the performance.

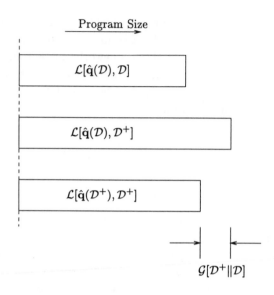

Fig. 7.6   Definition of Information Gain.

### 7.5.1.1  *Definition*

Let us define $\hat{\mathbf{q}}(\mathcal{D})$ as the minimal representation interpretation or the *best model* of data $\mathcal{D}$,

$$\hat{\mathbf{q}}(\mathcal{D}) \stackrel{\Delta}{=} \arg\min_{\mathbf{q}\in\mathbf{Q}} \mathcal{L}[\mathbf{q}, \mathcal{D}].$$

Suppose additional data is gathered, resulting in a larger data set $\mathcal{D}^+ \supset \mathcal{D}$. The **information gain** $\mathcal{G}[\mathcal{D}^+\|\mathcal{D}]$ due to the additional data $\mathcal{D}^+ - \mathcal{D}$ is defined as the *extra* bits needed to represent $\mathcal{D}^+$ using the best model for $\mathcal{D}$; thus

$$\boxed{\mathcal{G}[\mathcal{D}^+\|\mathcal{D}] \stackrel{\Delta}{=} \mathcal{L}[\hat{\mathbf{q}}(\mathcal{D}), \mathcal{D}^+] - \mathcal{L}[\hat{\mathbf{q}}(\mathcal{D}^+), \mathcal{D}^+]} \tag{7.38}$$

which is illustrated in Figure 7.6. Note that the information gain is always non-negative:

$$\mathcal{G}[\mathcal{D}^+\|\mathcal{D}] \geq 0,$$

since, from the definition of best model, $\mathcal{L}[\hat{\mathbf{q}}(\mathcal{D}^+), \mathcal{D}^+] \leq \mathcal{L}[\mathbf{q}, \mathcal{D}^+]$.

We define the "amount of information", in bits, obtained from an additional datum $\mathcal{D}'$ as the information gain due to $\mathcal{D}'$ over the original data set: $\mathcal{G}[\mathcal{D}, \mathcal{D}' \| \mathcal{D}]$. Intuitively, as additional data is gathered about the environment, we expect the information gain to converge to zero. In other words, as data samples $z_1, \ldots, z_N$ are acquired, we expect the sequence $\lim_{N \to \infty} \mathcal{G}[z_1, \ldots, z_N \| z_1, \ldots, z_{N-1}] \to 0$.

In general, when $\mathcal{D}^+$ and $\mathcal{D}$ are unrelated data sets; information gain $\mathcal{G}[\mathcal{D}^+ \| \mathcal{D}]$ can be regarded as a measure of the "information distance" or distortion between two data sets $\mathcal{D}$ and $\mathcal{D}^+$. In this sense, information gain is analogous to *discrimination* or directed divergence in statistical information theory [40].

### 7.5.1.2  *Adding More Sensors*

Let $\mathcal{D}^+$ denote the observed multisensor data, $\mathcal{D}^+ = \mathbf{Z}^{(1)}, \ldots, \mathbf{Z}^{(S)}$, and let $\mathcal{D}$ denote data from $s$th sensor, $\mathcal{D} = \mathbf{Z}^{(s)}, 1 \leq s \leq S$. Clearly $\mathcal{D}^+ \supset \mathcal{D}$, and therefore:

$$\mathcal{G}[\mathbf{Z}^{(1)}, \ldots, \mathbf{Z}^{(S)} \| \mathbf{Z}^{(s)}] \geq 0,$$

i.e., using an additional sensor gives non-negative information gain.

In other words, the best environment model found using all the multisensor data is *no worse* than that found using a single sensor's data. Thus, the interpretation (of all observed data) obtained using all the multisensor data is at least as good (in the sense of a smaller representation size) as the interpretation obtained using data from a single sensor alone (or a subset of sensors).

In practice, this property is meaningful only when all the sensors are observing the same environment.

### 7.5.2  *Redundancy in Multisensor Data*

We define the **redundancy** in multisensor data, denoted

$$\mathcal{J}[\omega^{(1)}, \ldots, \omega^{(S)}, \mathbf{Z}^{(1)}, \ldots, \mathbf{Z}^{(S)} \,|\, \mathbf{q}_{\mathrm{E}}]$$

as the *savings*, in bits, by jointly encoding the data, versus encoding each sensor's data separately and adding up the codelengths; thus

$$\mathcal{J}[\omega^{(1)},\dots,\omega^{(S)},\mathbf{Z}^{(1)},\dots,\mathbf{Z}^{(S)} \,|\, \mathbf{q}_{\mathrm{E}},\mathbf{Q}_{\mathrm{E}}]$$

$$\stackrel{\Delta}{=} \sum_{s=1}^{S} \mathcal{L}[\omega^{(s)},\mathbf{Z}^{(s)} \,|\, \mathbf{q}_{\mathrm{E}},\mathbf{Q}_{\mathrm{E}}] - \mathcal{L}[\omega^{(1)},\dots,\omega^{(S)},\mathbf{Z}^{(1)},\dots,\mathbf{Z}^{(S)} \,|\, \mathbf{q}_{\mathrm{E}},\mathbf{Q}_{\mathrm{E}}].$$

By definition, for independent sensors, the redundancy is zero (the converse may not be true).

Redundancy may be regarded as the amount of information the data sets $\mathbf{Z}^{(s)}$ replicate in each other, i.e., the number of bits saved by exploiting the sensor dependence in the joint encoding of data.

For sensors described by probabilistic uncertainty models, redundancy is the savings obtained using the joint probability distribution over the marginal distributions for each sensor; and it can be easily shown [40, Theorem 2.6.6] that the *average* value of redundancy is non-negative. For such sensors, the *average* value of redundancy identical to *mutual information* in statistical information theory [40].

On an average, exploiting the dependency in multisensor data leads to a smaller representation size than encoding the data independently. In practice, it may be easier to calculate the representation size of multisensor data assuming all sensors as independent, and then subtract a redundancy term or add a "conflict" term, based on some heuristics.

### 7.5.3   *Relative Confidence in an Interpretation*

Given a data set $\mathcal{D} = \mathbf{z}_1,\dots,\mathbf{z}_N$ arising from a model $\mathbf{q}^* \in \mathbf{Q}$, how many data samples, $N$, are needed to achieve a desired level of confidence?

We can define the *relative confidence* $\mathcal{C}(\mathbf{q},\mathcal{D}\,|\,\mathbf{Q})$, in a model $\mathbf{q}$ given a model library $\mathbf{Q}$ and observed data $\mathcal{D}$, by using the *Gibbs distribution*

$$\mathcal{C}(\mathbf{q},\mathcal{D}\,|\,\mathbf{Q}) \stackrel{\Delta}{=} \frac{2^{-\mathcal{L}[\mathbf{q},\mathcal{D}\,|\,\mathbf{Q}]}}{\sum_{\mathbf{q}\in\mathbf{Q}} 2^{-\mathcal{L}[\mathbf{q},\mathcal{D}\,|\,\mathbf{Q}]}} \tag{7.39}$$

For a given a data set $\mathcal{D}$, the relative confidence $\mathcal{C}(\mathbf{q},\mathcal{D}\,|\,\mathbf{Q})$ is a probability distribution in the model space $\mathbf{q} \in \mathbf{Q}$, and may be interpreted as the likelihood of a model.

As more data is acquired, the distribution $\mathcal{C}(\mathbf{q},\mathcal{D}\,|\,\mathbf{Q})$ should get narrower around the true model $\mathbf{q}^*$. In other words, as $N \to \infty$, we expect the relative confidence to approach a delta function

$$\lim_{N \to \infty} \mathcal{C}(\mathbf{q}, \mathbf{z}_1, \dots, \mathbf{z}_N \mid \mathbf{Q}) = \delta(\mathbf{q} - \mathbf{q}^*).$$

### 7.5.4 *Invariance to Sensor Transformations*

Given two sensors observing an environment, we say that they **equivalent**, if the data features observed by one sensor can be computed from the data features observed by the other sensor via a transformation mapping, and vice versa.

In general, not only do the observed data features undergo a transformation, but also the other components of the sensor model, including the uncertainty model and the constraint equation. An *invertible* transformation of the measurement space, $g : \Re^n \to \Re^n, \mathbf{z} \to \mathbf{z}' = g(\mathbf{z})$, just relabels the data features and transforms the constraint equation $\mathbf{h} \to \mathbf{h}'$, where $\mathbf{h}'(\mathbf{y}; \mathbf{z}') \triangleq \mathbf{h}(\mathbf{y}; g(\mathbf{z}))$.

Given equivalent sensors, a desirable property of any multisensor fusion criterion is that it should yield the same interpretation for either sensor. In other words, the criterion should have the same shape in "interpretation space", possibly shifted by a constant.

For the minimal representation size multisensor fusion criterion, the data from one sensor can be regenerated by encoding the data from the other sensor, and encoding the transformation describing the sensor mapping. For those transformation mappings that can be encoded in a fixed number of bits, independent of the observed data, the representation size remains invariant. Thus, for those sensors, the representation size expresses an intrinsic property of observed data, and is independent of the selection of measurement coordinate systems.

#### 7.5.4.1 *Invertible Linear Transformations*

Let us illustrate this with the case of an invertible linear transformation. For *invertible linear* transformations $g$, the constraint manifold doesn't change shape, and therefore $\mathcal{L}[\tilde{\mathbf{z}}' \mid \mathbf{y}] = \mathcal{L}[\tilde{\mathbf{z}} \mid \mathbf{y}]$.

The uncertainty models get transformed as follows. For a probabilistic uncertainty, Equation (6.8) is invariant, since the probability of $\Delta \mathbf{z}$ is the same as that of $\Delta \mathbf{z}'$. For an ellipsoidal uncertainty region, it is straightforward to show that $\Delta \mathbf{z}^{\mathrm{T}} A \Delta \mathbf{z} = \Delta \mathbf{z}'^{\mathrm{T}} A' \Delta \mathbf{z}'$, and therefore Equation (6.6) remains invariant.

Thus, the sensor representation size is invariant under linear non-singular (invertible) transformations of data.

## 7.6   Relationship to Classical Statistics

In this section we explore the relationship between the minimal representation size multisensor fusion approach and the classical approaches of Maximum A Posteriori (MAP) estimation and Maximum Likelihood (ML) estimation. For simplicity, we assume that only a single sensor is used to obtain data, from which the environment model and the data correspondence are to be estimated.

### 7.6.1   *MAP Estimation and Two-Part Encoding*

Let the probability distribution of observed data, given an environment model and a correspondence, be denoted by

$$p(\mathbf{Z} \,|\, \mathbf{q}_{\mathrm{E}}, \omega) = \prod_{i=1}^{N} p(\mathbf{z}_i \,|\, \mathbf{q}_{\mathrm{E}}, \omega)$$

where the data features are assumed to be independent. Assume the environment model and the correspondence are independent random variables,

$$p(\mathbf{q}_{\mathrm{E}}, \omega) = p(\mathbf{q}_{\mathrm{E}}) \cdot p(\omega),$$

where $p(\mathbf{q}_{\mathrm{E}})$ and $p(\omega)$ are the *a priori* probabilities on the environment model and the correspondence respectively. Let us define the probability of observing a correspondence $\omega$ as the number of ways of arranging $N$ labels among $M + 1$ categories so that there are exactly $N_j$ labels in category $j$, divided by the total number of ways of assigning $N$ labels; thus

$$p(\omega) \triangleq \frac{N!/N_0! \dots N_M!}{N!} \approx \frac{1}{N_0^{N_0} \dots N_M^{N_M}} = \frac{f_0^{N_0} \dots f_M^{N_M}}{N^N}$$

using Stirlings approximation (see page 104). Using Bayes rule [151], the *a posteriori* probability of an environment model $\mathbf{q}_{\mathrm{E}}$ and a correspondence $\omega$

is given by

$$p(\mathbf{q}_E, \omega \mid \mathbf{Z}) = \frac{\prod_{i=1}^{N} p(\mathbf{z}_i \mid \mathbf{q}_E, \omega) \cdot p(\mathbf{q}_E, \omega)}{p(\mathbf{Z})}$$

The MAP estimation criterion is obtained by maximizing the posteriori probability $p(\mathbf{q}_E, \omega \mid \mathbf{Z})$, and is equivalent to minimizing its negative logarithm; thus we minimize

$$-\log_2 p(\mathbf{q}_E, \omega) - \log_2 \prod_{i=1}^{N} p(\mathbf{z}_i \mid \mathbf{q}_E, \omega)$$

$$= -\log_2 p(\mathbf{q}_E) - \sum_{j=0}^{M} N_j \cdot \log_2 f_j - \sum_{i=1}^{N} \log_2 p(\mathbf{z}_i \mid \mathbf{y}_{\omega(i)})$$

where the $p(\mathbf{Z})$ and the $N \cdot \log_2 N$ terms were dropped since they are constant for a given data set.

Comparing this expression with the *two-part* representation size expression in Table 7.1 and noting that $p(\mathbf{z}_i \mid \mathbf{y}_{\omega(i)}) \equiv 2^{-\mathcal{L}[\mathbf{z}_i \mid \mathbf{y}_{\omega(i)}]}$, we see that their *last two terms are identical*. The $-\log_2 p(\mathbf{q}_E)$ term is the equivalent of the $\mathcal{L}[\mathbf{q}_E \mid \mathbf{Q}_E]$ term in Equation (7.10), since we considered only a single sensor.

There is no equivalent of the algorithm size term $\mathcal{L}[\mathcal{A} \mid \mathbf{q}_E, \mathbf{Q}_E]$ in the MAP estimation criterion.

### 7.6.2 *ML Estimation and Mixture Encoding*

A ML criterion can be developed in a similar manner. In the ML approach, the environment model $\mathbf{q}_E$ and the correspondence $\omega$ are no longer regarded as random variables; instead they are regarded as parameters to the probability distribution generating the observed data, denoted by

$$p(\mathbf{Z}; \mathbf{q}_E, \omega) = \prod_{i=1}^{N} p(\mathbf{z}_i; \mathbf{q}_E, \omega),$$

assuming independent data features. We may regard $\mathbf{z}$ as a mixture of $M + 1$ random variables,

$$p(\mathbf{z}_i; \mathbf{q}_E, \omega) = \sum_{j=0}^{M} f_j \cdot p(\mathbf{z}_i \mid \mathbf{y}_j)$$

where $f_j \triangleq N_j/N$. Combining the two expressions, the *likelihood* of observed data is given by

$$p(\mathbf{Z}; \mathbf{q_E}, \omega) = \prod_{i=1}^{N} \sum_{j=0}^{M} f_j \cdot p(\mathbf{z}_i \,|\, \mathbf{y}_j).$$

The ML criterion is obtained by maximizing the likelihood, or minimizing its negative logarithm; thus we minimize

$$- \log_2 \prod_{i=1}^{N} \sum_{j=0}^{M} f_j \cdot p(\mathbf{z}_i \,|\, \mathbf{y}_j) = - \sum_{i=1}^{N} \log_2 (\sum_{j=0}^{M} f_j \cdot p(\mathbf{z}_i \,|\, \mathbf{y}_j)).$$

Comparing this expression with the *mixture* representation size expression in Table 7.1 and noting that $p(\mathbf{z}_i \,|\, \mathbf{y}_j) \equiv 2^{-\mathcal{L}[\mathbf{z}_i \,|\, \mathbf{y}_j]}$, we see that its *last term is identical* to the ML criterion above.

There is no equivalent for the algorithm size term $\mathcal{L}[\mathcal{A} \,|\, \mathbf{q_E}, \mathbf{Q_E}]$, or for the model complexity term $\mathcal{L}[\mathbf{q_E} \,|\, \mathbf{Q_E}]$ in the ML criterion.

### 7.6.3 *Remarks*

From the discussion in the preceding paragraphs, it is clear that the MAP and the ML criterion can be regarded as special cases of the minimal representation size criterion.

In particular, the algorithm representation size $\mathcal{L}[\mathcal{A} \,|\, \mathbf{q_E}]$ is absent in the MAP and ML estimation criteria. This term plays a significant role in discriminating between different model classes, especially when the number of model features depend on the model parameters. The model representation size term $\mathcal{L}[\mathbf{q_E} \,|\, \mathbf{Q_E}]$ in Equation (7.10) also plays a significant role in model structure discrimination and selection; it is absent in the ML estimation criterion.

# Chapter 8

# Multisensor Fusion Search Algorithms

## 8.1 Introduction

In the minimal representation size multisensor fusion framework, the *best interpretation* of observed multisensor data is found by minimizing the multisensor representation size criterion (Equation 7.10 on page 97). A *multisensor fusion algorithm* searches for the minimal representation size interpretation, in the space of all possible environment models and legal correspondences, and is illustrated in Figure 8.1. In general, this search space can be quite large and exhaustive search quickly becomes infeasible for practical problems. As noted in Section 6.2, the space of discrete correspondences grows exponentially with the number of observed data features. The environment model parameter space is typically a continuous subspace of $\Re^D$.

In this chapter we discuss several search algorithms which can be applied to multisensor fusion problems within the minimal representation size framework. There are three basic components of these search problems:

(1) **Environment model instantiation:** An environment model structure, chosen from the library, must be instantiated with specific parameters. Thus, $q_E = \Xi(\theta)$, where the structure, $\Xi(\cdot)$, and the parameters, $\theta$, must be somehow instantiated. Methods for instantiating environment models are discussed in Section 8.2.

(2) **Correspondence Computation:** The minimal representation size correspondences must be computed given an instantiated environment model. Algorithms for finding the minimal representation size correspondence are described in Section 8.3.

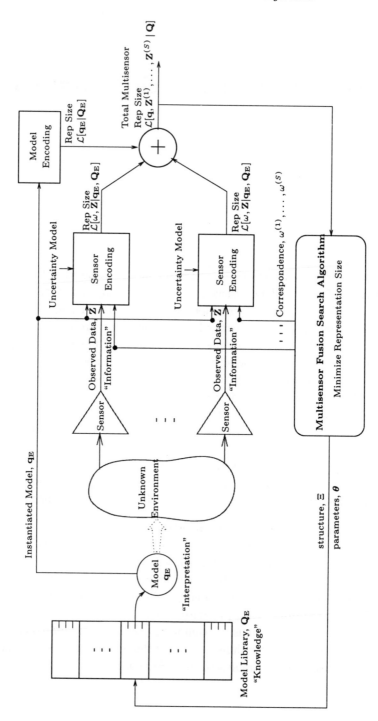

Fig. 8.1 Minimal representation multisensor fusion and model selection.

(3) **Search engine:** A search engine *systematically* instantiates environment models, computing the minimal representation size correspondences for each; in order to find a good interpretation in the space of possibilities. In this work we will describe two classes of search algorithms: a polynomial time hypothesize and test algorithm described in Section 8.4, and a class of evolutionary algorithms described in Section 8.5.

These two types of algorithms have been implemented in a "generic" form, that is, no additional heuristics or specific problem knowledge has been added. This approach permits a more objective assessment of the minimal representation size criterion for multisensor fusion. However, in practice, problem specific heuristics might be added to improve the efficiency of implementation for a particular problem domain. For example, in object recognition problems, typical heuristics might include: **(a)** pose space clustering to reduce the search space (Section 4.2.1.1), **(b)** using transform space geometry to prune pose parameters (Section 4.2.1.2), **(c)** pruning matches based on relations which must be satisfied by both data and model features (Section 4.2.2.6), and **(d)** using a fixed size subset of data features for generating pose hypothesis (Section 4.2.1.3).

## 8.2 Instantiating Environment Models

Given a model structure, $\Xi(\cdot)$, in general, the associated parameters, $\theta$, are drawn from a $D$-dimensional set, $\Theta \in \Re^D$. We discuss two ways of choosing $\theta$, to instantiate environment models, $\mathbf{q}_E = \Xi(\theta)$.

### 8.2.1 *Random*

The environment model parameters can be drawn randomly from the set of possible values; thus $\theta \in \Theta$. Typically, a uniform distribution is used, unless a stronger *a priori* distribution is available. This method is useful mainly for the evolutionary algorithms discussed in Section 8.5, which maintain a population of environment models.

### 8.2.2    *Constraining Data Feature Sets*

Constraining Data Feature Sets (CDFS) are a practical heuristic to instantiate environment model parameters (for given a model structure) based on the observed sensor data. A CDFS is defined as a *smallest* subset of data features, sufficient to determine all the model parameters $\theta$ for a given model structure $\Xi(\cdot)$. We pair $K$ observed data features, $z_{i_1}, \ldots, z_{i_K}$, with $K$ *arbitrary* model features, $y_{j_1}, \ldots, y_{j_K}$, and solve the resulting system of constraint equations to instantiate the environment model parameters:

$$\mathbf{h}(\mathbf{y}_{j_1}; \mathbf{z}_{i_1}) \;=\; 0$$

$$\vdots$$

$$\mathbf{h}(\mathbf{y}_{j_K}; \mathbf{z}_{i_K}) \;=\; 0$$

Therefore, a CDFS is that smallest set of observed data features, which allows determination of parameters $\theta$ by solving the associated system of constraint equations. CDFS could be comprised of heterogeneous data features, using constraint equations from multiple sensors; thus several types of CDFS may be available for a given multisensor fusion problem.

A CDFS system of constraint equations may be exactly or *partially* solved to hypothesize environment model parameters. This method of instantiating parameters *exploits* the observed multisensor data, and samples a "promising" subspace of the parameter space $\Theta$. This heuristic is particularly suitable for use with ellipsoidal accuracy and precision sensor uncertainty models, since they lead to an error residual encoding which selects model parameters which favor partial "perfect" fits, where some of the error residuals are zero or very small, as discussed in Section 7.4.2.1.

For each arbitrary pairing of data and model features, a CDFS can generate several parameter vectors, $\theta \in \Theta$. For a finite system of constraint equations, the number of such pairings is *polynomial* in the number of data and model features. Therefore, the CDFS parameters sample a *polynomial size subspace* of the $D$-dimensional parameter space.

## 8.3   Finding the Best Correspondence

Given an instantiated environment model, $\mathbf{q}_{\text{E}} = \Xi(\theta)$, the minimal representation size correspondence must be computed *for each sensor*, as prescribed by Equations (7.6) and (7.7).

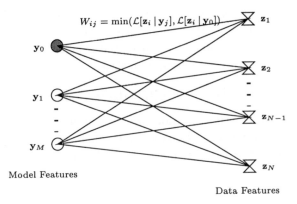

Fig. 8.2  A weighted graph is constructed for a sensor, by pairing each data feature with each model feature, and assigning a weight $W_{ij}$.

In Section 8.3.1, we discuss algorithms to find the discrete correspondence mapping, $\omega$, used in a two-part encoding scheme. In Section 8.3.2, we outline methods to find the relative frequency vector $\mathbf{f}$, used in a mixture encoding scheme.

### 8.3.1  *Discrete Correspondence Mapping*

To search for the minimal representation size discrete correspondence mapping, $\omega$, we simplify the two-part representation size expression (Table 7.1) by ignoring the $\mathcal{L}[\omega \,|\, \mathcal{A}, \mathbf{q}_E, \mathbf{Q}_E]$ term; thus

$$\mathcal{L}[\omega, \mathbf{Z} \,|\, \mathcal{A}, \mathbf{q}_E, \mathbf{Q}_E] \approx \sum_{i=1}^{N} \mathcal{L}[\mathbf{z} \,|\, \mathbf{y}_{\omega(i)}]. \qquad (8.1)$$

Note that this simplification is equivalent to using a fixed-length encoding scheme described in Section 7.3.2.2. We use this simplification only to compute the best correspondence (i.e., only to simplify the search); the original representation size expression from Table 7.1 is used to evaluate the total representation size of the resulting interpretation (which is used for model selection).

The simplification of Equation (8.1) allows us to translate the correspondence search problem, into a *minimum weight graph matching problem*, as follows. As illustrated in Figure 8.2, a weighted graph is constructed with

$M + 1$ model features comprising the left nodes, and $N$ data features comprising the right nodes. Each model feature is connected to every data feature by an arc with weight equal to the smaller of the modeled and unmodeled data features representation sizes:

$$W_{ij} = \min(\mathcal{L}[\mathbf{z}_i \,|\, \mathbf{y}_j], \mathcal{L}[\mathbf{z}_i \,|\, \mathbf{y}_0]),$$

since the unmodeled representation size gives an upper bound on the data feature representation size (Equation 7.18 on page 103). This graph is represented by a $N \times (M + 1)$ matrix with the $ij$-th element equal to the weight $W_{ij}$.

**Many-to-One Correspondence** In the case of *many-to-one* correspondence, $\omega(i)$ is given by the index of the minimum element in $i$-th row of the graph matrix. When several elements have the same minimum value, the smallest index is chosen.

This procedure takes $O(M)$ steps per row to find the row minimum, and therefore a total of $O(N \cdot M)$ time steps to find $\omega$.

**One-to-One Correspondence** To enforce the one-to-one constraint (Section 6.2.2), the graph of Figure 8.2 is modified into a *complete bipartite graph*, by making the number of left nodes equal the number of right nodes, as follows. If $M + 1 < N$, add $N - M - 1$ "extra" nodes to the set of model feature nodes on the left side. Connect each extra left node to every data feature node using $N$ arcs, each with a weight $\mathcal{L}[\mathbf{z}_i \,|\, \mathbf{y}_0]$. If, on the other hand, $N < M + 1$, add $M + 1 - N$ "extra" nodes to the set of data feature nodes on the right side. Connect each extra right node to every model feature node using $M + 1$ arcs, each with weight 0.

Such a graph of representation sizes is a complete bipartite graph. The minimal representation correspondence is defined by choosing $\max(M + 1, N)$ *distinct* arcs of the modified graph such that the sum of the arc weights is a minimum. For data feature $i$, a modeled correspondence is indicated by an arc with weight less than $\mathcal{L}[\mathbf{z}_i \,|\, \mathbf{y}_0]$; an arc with weight $\mathcal{L}[\mathbf{z}_i \,|\, \mathbf{y}_0]$ indicates that it is unmodeled. The sum of the arc weights is the representation size of the minimal representation match.

Computing the minimal representation correspondence in this manner amounts to solving the problem of finding a *minimum weight linear assignment* of the left nodes to the right nodes, also known as the minimum weight bipartite graph matching problem. It has been studied extensively in the literature [67] and a number of efficient algorithms exist. We use one such assignment algorithm [22] that runs in $O(M \cdot N \cdot \min(M, N))$ time steps.

## 8.3.2 *Relative Frequency Vector*

To search for the minimal representation size relative frequency vector, $\mathbf{f}$, note that the mixture representation size expression (Table 7.1):

$$\mathcal{L}[\mathbf{f}, \mathbf{Z} \mid \mathcal{A}, \mathbf{q}_{\text{E}}, \mathbf{Q}_{\text{E}}] \equiv \mathcal{L}[\omega, \mathbf{Z} \mid \mathcal{A}, \mathbf{q}_{\text{E}}, \mathbf{Q}_{\text{E}}] = -\log_2 \{ \sum_{j=0}^{M} f_j \cdot 2^{-\mathcal{L}[\mathbf{z} \mid \mathbf{y}_j]} \}, \quad (8.2)$$

is strictly convex $\cup$ in $\mathbf{f}$, as discussed in Appendix E. Therefore, the best relative frequency vector, $\hat{\mathbf{f}}$, can be computed using any standard *convex optimization* method [197], subject to the constraints on $\mathbf{f}$.

The resulting $(M + 1)$-dimensional constrained convex optimization is independent of the number of observed data features $N$.

**Many-to-One Correspondence** For a many-to-one correspondence, the relative frequency vector, $\mathbf{f}$, must satisfy the following *unit simplex* constraint:

$$\mathbf{f}: \quad f_j \in [0, 1], \qquad \sum_{j=0}^{M} f_j = 1,$$

where, $j = 0, 1, \ldots, M$.

**One-to-One Correspondence** For one-to-one correspondence, the relative frequency vector, $\mathbf{f}$, is subject to the constraints described in Section 7.3.2.2, which are summarized below:

$$\mathbf{f}: \quad f_0 \in [1 - \frac{\min(M, N)}{N}, 1], \qquad f_{j \neq 0} \in \{0, \frac{1}{N}\}, \qquad \sum_{j=0}^{M} f_j = 1,$$

where, $j = 0, 1, \ldots, M$.

## 8.4   Hypothesize and Test Algorithms

A *hypothesize and test* algorithm which uses CDFS to instantiate environment model parameters, and finds the best correspondence using the algorithms described in Section 8.3, is summarized in Figure 8.3.

(1) For each environment model structure $\Xi$ in the library, find its "best" interpretation that minimizes the multisensor representation size:

    (a) Select a CDFS and use it to hypothesize model parameters $\boldsymbol{\theta}$.

    (b) For each hypothesized model parameter $\boldsymbol{\theta}$, evaluate the minimal representation size $\mathcal{L}[\mathbf{Z}^{(s)} \mid \mathbf{q_E} = \Xi(\boldsymbol{\theta})]$ (Equation 7.6) of observed multisensor data. Use the algorithms discussed in Section 8.3 to search through the correspondence space, and find the minimal representation size correspondences, $\hat{\omega}^{(s)}(\boldsymbol{\theta})$, for each sensor.

    (c) Repeat until all possible CDFS combinations are exhausted. Keep the smallest representation size parameter $\hat{\boldsymbol{\theta}}$ and the data correspondences, $\hat{\omega}^{(s)}(\hat{\boldsymbol{\theta}})$, for each sensor (Equation 7.7).

(2) For the minimal representation interpretation thus found, evaluate the total representation size $\mathcal{L}[\mathbf{q_E} \mid \mathbf{Q_E}] + \sum_{s=1}^{S} \mathcal{L}[\mathbf{Z}^{(s)} \mid \mathbf{q_E}, \mathbf{Q_E}]$ (Equation 7.5).

(3) Repeat for all environment model structures. Keep the interpretation with the smallest total representation size—this give the estimated minimal representation environment model $\hat{\mathbf{q}}_E = \hat{\Xi}(\hat{\boldsymbol{\theta}})$, and the associated correspondences $\hat{\omega}^{(s)}(\hat{\boldsymbol{\theta}})$, $s = 1, \ldots, S$, for all the sensors.

Such an algorithm operates in *polynomial time*, since the sampled subspace of CDFS parameters in Step 1c is of polynomial size, and in Step 1b, the correspondence space can be searched in polynomial time as discussed in Section 8.3.

In practice, domain specific heuristics and constraints can be exploited to speed up the search, and it may be possible to design more efficient search algorithms, specialized to particular problem domains.

## 8.5 Evolution Programs

The general principles for harnessing simulated natural evolution to solve search problems, have been firmly established by the nearly three decades of research on *genetic algorithms* (GA) [66], *genetic programming* (GP) [114, 115], *evolutionary strategies* (ES) [7, 176], and *evolutionary programming* (EP) [59]. These principles define a general *paradigm* for mimicking natural evolution on a population of individuals, and form the basis of a general class of evolutionary algorithms [8, 136].

A good reference on the general principles of evolutionary algorithms, is the book by Michalewicz [136]. Following that book's suggestion, we use the term **evolution program** to refer to the application of evolutionary principles to domain specific representations (data structures), to solve particular problems. Another good reference, especially for parameter optimization, is the overview of evolutionary algorithms by Bäck and Schwefel [8].

Evolution programs are often quite practical for minimizing cost functions with multiple local minima and complex surfaces (Equation 7.10), but usually offer no analytical bounds on convergence rates or the guarantee of finding the global minimum. In Section 8.5.1, we describe an evolution program template which encapsulates the general principles of simulating natural evolution. This template encompasses many of the established evolutionary algorithms (such as GA, ES, EP), and can be used to describe and develop new ones. In Section 8.5.3, we describe how such evolution programs may be used for minimal representation size multisensor fusion and model selection problems. For the multisensor object recognition applications, described in Part 3, we have found one recently developed evolutionary algorithm, namely *differential evolution* (DE) [190], to be particularly suitable. In Section 8.5.2, we describe this algorithm in more detail.

### 8.5.1 *An Evolution Program Template*

Evolution programs search by maintaining, at any time $t$, a *population*, $\mathcal{P}(t)$, of $P$ individuals, and applying the rules of natural selection to generate the population at time $t + 1$, as shown in Figure 8.4. Each individual in the population, describes a specific instance of the search problem $\mathbf{p}$, represented by a data structure $\mathcal{S}(\mathbf{p})$; thus

$$\mathcal{P}(t) = \{\mathcal{S}(\mathbf{p}_1), \ldots, \mathcal{S}(\mathbf{p}_P)\}$$

The choice of the **representation** $\mathcal{S}(\cdot)$ is of critical importance in successfully applying the evolutionary paradigm to a given search problem.

Closely linked with the choice of representation, is the choice of a **fitness** function, $\Phi(\mathcal{S}(\mathbf{p})) \longrightarrow \Re$, which assigns a numerical "merit" to each individual in the population. In general, the fitness function, $\Phi(\cdot)$, and the representation, $\mathcal{S}(\cdot)$, should be chosen such that *problems which are "close" in the* problem space *are also close in the* representation space. Thus, if the problem space "distance" between problems $\mathbf{p}_i$ and $\mathbf{p}_j$ is small, the distance between their corresponding representations, $\mathcal{S}(\mathbf{p}_i)$ and $\mathcal{S}(\mathbf{p}_j)$, given by the difference in fitness values $|\Phi(\mathcal{S}(\mathbf{p}_i)) - \Phi(\mathcal{S}(\mathbf{p}_j))|$ should also be small. Defining a notion of "distance" on the problem space is often difficult, and this problem formulation is often critical in developing effective evolution programs for new problem domains. For parameter optimization problems, the problem domain distance is well-defined by the objective function, and therefore the choice of the representation and the fitness function becomes somewhat easier.

The $P$ individuals in the *parent* population, $\mathcal{P}(t)$, undergo **reproduction** to produce an *offspring* population, $\mathcal{P}''(t)$, of $R$ individuals (Figure 8.4). Reproduction is accomplished by means of crossover and mutation operators which transform the parent population as follows. A **crossover** operator, controlled by the parameters $\nu_c$, recombines two or more individuals in the parent population, to produce an individual in the intermediate population $\mathcal{P}'(t)$ of $R$ individuals:

$$\mathcal{S}(\mathbf{p}) \times \ldots \times \mathcal{S}(\mathbf{p}) \xrightarrow[\text{crossover}]{\nu_c} \mathcal{S}(\mathbf{p}).$$

A **mutation** operator, controlled by the parameters $\nu_m$, transforms each intermediate individual, to produce an individual in the offspring population $\mathcal{P}''(t)$ of $R$ members:

$$\mathcal{S}(\mathbf{p}) \xrightarrow[\text{mutation}]{\nu_m} \mathcal{S}(\mathbf{p}).$$

The choice of the crossover and the mutation operators is also critically important to the successful use of an evolution program, and depends on the chosen representation. These operators should be *designed* to operate on the chosen representation $\mathcal{S}(\cdot)$ so that they can **(a)** *exploit* the "promising" areas of the search space implicit in the parent population, and **(b)** *explore* new areas of the search space which have not yet been visited, in generating new individuals. The crossover and mutation operator parameters, namely $\nu_c$ and $\nu_m$, control the output of the reproduction stage, and should be chosen to balance the need for maintaining *population diversity* and *selective pressure*, in producing the offspring population.

The offsprings and the parents compete for *survival* according to the rules of natural selection. A *selection* operator, controlled by parameters $\nu_s$, chooses $P$ survivors from the pool of $R$ offsprings and (optionally) the $P$ parents, based on their fitness values:

$$\mathcal{P}''(t) \cup \mathcal{P}(t) \xrightarrow[\text{select}]{\nu_s} \mathcal{P}(t+1)$$

These $P$ survivors comprise the parent population of the next generation, $\mathcal{P}(t+1)$, as shown in Figure 8.4. Numerous selection operators, many inspired by natural evolution, have been documented in the literature. One selection operator may perform better than another, given a specific problem representation and associated reproduction operators. However, clear principles for choosing among the various selection operators have not yet emerged.

The cycle of reproduction, evaluation, and selection is repeated until some **termination condition** is met (Figure 8.4). The successive application of reproduction and selection operators on a population of individuals, leads to new populations which are "more fit" (on an average), than the preceding ones. The emerging population may contain several clusters in the representation space, which typically shrink with each successive generation. For a parameter optimization problem, each cluster may describe a local optimum. The best local optimum, thus found, is often the global optimum to a search problem.

The effectiveness of evolution program often depends on the choice of the **initial population**. Often, the initial population is drawn randomly from the search space. Problem specific heuristics may be useful in choosing an initial population which exploits the search space more effectively.

### 8.5.2 *Differential Evolution*

The *differential evolution* (DE) algorithm [156, 189, 190] has been successfully used for solving several difficult parameter optimization problems [190]. We describe it in terms of the evolution program template presented above.

### 8.5.2.1  Representation

The DE uses a floating point string to represent a parameter vector. Thus, for an optimization problem, described by a $D$-dimensional parameter vector $\mathbf{p}$:

$$\mathcal{S}(\mathbf{p}) = \mathbf{p} = [\mathbf{p}(1) \quad \cdots \quad \mathbf{p}(D)]^{\mathrm{T}}$$

where, $\mathcal{S}(\mathbf{p})$ is an array of $D$ *reals*.

### 8.5.2.2  Fitness Function

The fitness function is chosen to be the objective function to be *minimized*:

$$\Phi(\mathcal{S}(\mathbf{p})) = \Phi(\mathbf{p}) = \text{Objective}(\mathbf{p}).$$

### 8.5.2.3  Reproduction

In DE the size of the offspring population is equal to that of the parent population, $R = P$.

The **crossover** operator utilizes the differentials in the parent population, to generate new offspring. It operates on each member of the parent population as follows. A *differential perturbation* is applied to $i$-th individual, $\mathbf{p}_i$, in the parent population, to generate a new individual, $\mathbf{p}'_i$:

$$\mathbf{p}'_i(d) = \begin{cases} \gamma \cdot \mathbf{p}_{\text{best}}(d) + (1-\gamma) \cdot \mathbf{p}_{i_a}(d) + \mathrm{F} \cdot \overbrace{\sum_{k=1}^{K} [\mathbf{p}_{i_b^k}(d) - \mathbf{p}_{i_c^k}(d)]}^{\text{differential perturbation}} & d \in \mathcal{D} \\ \mathbf{p}(d) & \text{otherwise} \end{cases}$$

where $\mathbf{p}_{\text{best}}$ is the best individual in parent population, $\gamma \in [0,1]$ is the *greediness* of the operator, $K$ is the number of *differentials* used to generate the perturbation, $\mathrm{F}$ is a factor used to *scale* the differentials, $\mathcal{D}$ is the set of element *indices* that undergo perturbation, and $i_a, i_b^k, i_c^k$ are randomly selected *mutually distinct* individuals in the parent population, distinct from the current individual, $i$. The DE crossover operator is illustrated in Figure 8.5.

The greediness, $\gamma$, sets the actual parameter value which undergoes perturbation. For $\gamma = 0$, a randomly chosen individual in the parent population, namely $i_a$, is perturbed. For $\gamma = 1$, the best member of the parent population is perturbed. For intermediate values of $\gamma$, a weighted sum of the best individual and a randomly chosen individual in the parent population is used. Typically one or two differentials are used, i.e., $K \in \{1, 2\}$, and the scale factor $F \in [0, 2]$.

The set of parameter vector element indices, $\mathcal{D}$, which undergo perturbation, is controlled by the crossover probability, $CR \in [0, 1]$, and chosen according to one of the following rules, where $\mathcal{U}(0, 1)$ is a uniform random number generator in $[0, 1)$:

- *Exponential crossover:*
  $\mathcal{D} \leftarrow \emptyset$
  $l = $ randomly chosen in $0, \ldots, D - 1$
  do
  $\qquad \mathcal{D} \leftarrow \mathcal{D} \cup \{l + 1\}$
  $\qquad l \leftarrow l + 1 \mod D$
  while $\mathcal{U}(0, 1) < CR$ and $|\mathcal{D}| < D$

  The parameter vector is viewed as a circular array, and this scheme selects a *substring* of $|\mathcal{D}|$ adjacent elements, to be perturbed.

- *Binomial crossover:*
  $l^* = $ randomly chosen in $1, \ldots, D$
  $\mathcal{D} \leftarrow \mathcal{D} \cup \{l^*\}$
  for each $l \in \{1, \ldots, D\}$ and $l \neq l^*$
  $\qquad$ if $\mathcal{U}(0, 1) < CR$
  $\qquad\qquad \mathcal{D} \leftarrow \mathcal{D} \cup \{l\}$

  This scheme selects $|\mathcal{D}|$ elements to be perturbed, uniformly distributed throughout the parameter vector (by tossing a biased coin for each of the $D$ places).

Note that $\mathcal{D}$ contains at least one element (exactly one element when $CR = 0$), and at most $D$ elements (exactly $D$ elements when $CR = 1$).

The greediness $\gamma$, the number of differentials $K$, the scale factor $F$, and the crossover probability, $CR$, constitute the parameters of the DE crossover operator:

$$\nu_c = < \gamma, K, F, CR >,$$

and must be chosen appropriately of a each specific type of minimization problem.

There is *no mutation* operator in the standard DE; thus

$$\mathbf{p}_i'' = \mathbf{p}_i'$$

where $i = 1, \ldots, R$.

Various flavors of DE are described using the notation:

$$\mathrm{DE}/\gamma/K.$$

The commonly used varieties are DE/0/1 and DE/1/2, which are often written as DE/rand/1 and DE/best/2 respectively. The exponential crossover is the one commonly used.

### 8.5.2.4  *Selection*

DE uses deterministic selection, where each offspring competes with its parent, and survives only if its fitness is better:

$$\mathbf{p}_i^{(t+1)} = \begin{cases} \mathbf{p}_i''^{(t)} & \text{if } \Phi(\mathbf{p}_i''^{(t)}) < \Phi(\mathbf{p}_i^{(t)}) \\ \mathbf{p}_i^{(t)} & \text{otherwise.} \end{cases}$$

### 8.5.2.5  *Initialization*

As the population converges, the population cluster shrinks, and therefore the magnitude of the differential perturbation also diminishes—thus the differential perturbations are automatically scaled to a size appropriate for the current population. However, this means that that the initial population must be chosen to sample the entire parameter space—otherwise the differentials will not be diverse enough to explore some parts of the parameter search space.

In practice, the initial parameters are drawn randomly, according to a uniform distribution in the parameter space.

### 8.5.2.6  *Choosing the Differential Evolution Search Parameters*

The values for the DE search parameters are selected by "trial and error". According to the rules of thumb described in the literature [156, 190], reasonable starting values are $P = 10D$, F $= 0.5$, and CR $= 0.5$. These values are adjusted to achieve consistent convergence in a reasonable amount of time. The number of differentials $K$ and the greediness $\gamma$ are also selected based on the speed of convergence; often $K = 2$ and $\gamma = 1$ give the fastest convergence [156].

From our practical experience with DE, increasing $P$ only helps to a certain extent. If $P$ is very small, the population does not represent a good sampling of the search space, and the space of possible perturbations is not diverse enough to explore promising directions. Making $P$ very large does not seem to help convergence either, since the selection pressure diminishes and the information contained in the current population cannot be efficiently exploited (too many "poor" individuals dilute the effect of "good" individuals).

An informal analysis follows. For a $K$ differential DE, the number of possible differential perturbations is given by (Figure 8.5):

$$\binom{P}{2K} \cdot 2K! \approx O(P^{2K}).$$

The number of possible perturbations determines the number of directions that can be explored at each generation. "Good" perturbations direct the population towards the minima in the search space, and the number of "good" perturbations depends on the topography of the search space. The ratio of the "good" perturbations to the total number of possible perturbations determines the effectiveness of the evolutionary search. As $P$ increases, this ratio diminishes, thus reducing the selection pressure in the search, and taking longer to converge. Conversely, if $P$ is too small, the population is not diverse enough to generate perturbations which effectively explore the search space.

The scale factor F also affects the performance of the search. It should be small enough so that the differentials explore the "tight" valleys, and it should be large enough to maintain a diverse population that can explore new areas of the search landscape. As the size of the differential perturbation space increases with increasing $P$, the weight F should decrease to maintain the selection pressure. The choice of $P$ and F determines the ability of the DE to find a solution—these parameters ought to be selected appropriately for a class of problems, after considering the topography of the search space. Perhaps F should be adapted, as the population evolves, to adjust to the specific instance of the search space. Large values of $P$ and F often result in premature convergence.

The crossover probability CR controls the number of parameter vector

elements that get replaced in the offspring due to the differential perturbation. In a multidimensional search space, this controls the number of
dimensions that get changed between the parent and the offspring. Decreasing CR lowers this number (exactly one dimension is updated for CR = 0)
increasing the search robustness, but taking longer. Increasing CR increases this number (the entire parent is replaced for CR = 1), and often results
in faster convergence.

### 8.5.3  Mapping Multisensor Fusion and Model Selection Problems

The minimal representation size multisensor fusion and model selection
criterion (Equation 7.10) is a "parameter optimization" problem, where
the unknowns consist of the environment model structure, the environment
model parameters, and the data correspondences (Figure 7.1):

$$\mathbf{p} \overset{\Delta}{=} < \Xi(\boldsymbol{\theta}), \omega^{(1)}, \ldots, \omega^{(S)} > .$$

In this section we outline a generic template for mapping the abstract
multisensor fusion and model selection problem into evolution programs.

#### 8.5.3.1  *General Considerations*

In this section we discuss a few general patterns that are useful in devising
evolution programs for minimal representation size multisensor fusion and
model selection problems.

**Environment Model Representations** The environment model parameters are often represented as vectors of floating point numbers
in $\mathcal{S}(\mathbf{p})$. The environment model parameters are drawn from a set
$\theta \in \Theta$:

$$\theta_d \in [L(d), U(d)]$$

where $L(d)$ and $U(d)$ are the lower and upper bounds on the $d$-th
element of the $D$-dimensional parameter vector. The environment
model parameters may be subject to additional constraints.
Sometimes, a representation can be chosen such that the new offspring generated by the reproduction operators, always satisfy the
environment model parameter constraints. Often, this may not be

possible, and in those cases, we assign a very "poor" fitness to these individuals (i.e., we assign a representation size of $\infty$).

The environment model structure labels (or ids) may also be encoded as an additional element in $S(\mathbf{p})$, especially for pure model class selection problems, for which the number of parameters is the same for all environment model structures.

An alternative representation of the environment model parameters can be based on the notion of CDFS (Section 8.2.2). The representation, $S(\mathbf{p})$, includes an environment model structure label, and pairs of data and model features *indices* which constitute a CDFS. *Specialized* reproduction operators must be devised so that invalid CDFS index combinations are not generated.

**Correspondence Representations** Often, the problem representation consists of only the environment model structure and parameters; the correspondences are computed directly using the algorithms in Section 8.3. Sometimes (especially when the number of model or data feature is large), it is desirable to also perform this search using an evolutionary algorithm. In those cases, several alternative correspondence representations may be used in $S(\mathbf{p})$.

**Discrete Correspondence Mapping** A discrete correspondence, $\omega$, may be represented as a binary or floating point string of $N$ numbers, whose elements take values in the range $0, \ldots, M$. If a floating point string is used, each element can be rounded to the nearest integer (for evaluation purposes only).

A *one-to-one correspondence*, must satisfy additional constraints on the allowed values (Section 6.2.2). This can be handled by assigning a very "poor" fitness to such individuals (i.e., we assign a representation size of $\infty$). Alternatively, *specialized* reproduction operators may be developed, which always produce legal a correspondence in the offspring. This latter approach was adopted by Ravichandran [158], to devise a specialized GA for solving matching problems.

**Relative Frequency Vector** A relative frequency vector, $\mathbf{f}$, may be represented as a floating point string of $M + 1$ numbers, whose elements take values in the range $[0, 1]$, and must sum up to 1. For a *one-to-one correspondence*, the relative frequency vector must satisfy additional constraints (Section 8.3.2). Individuals which violate the constraint requirements can be assigned a very "poor" fitness (i.e., a representation size of $\infty$). Alternatively, *specialized* reproduction operators may be devised, which guarantee a legal offspring.

**Fitness Evaluation** The fitness function is the total multisensor representation size (Equation 7.10):

$$\Phi(\mathcal{S}(\mathbf{p})) = \mathcal{L}[\mathbf{q}, \mathbf{Z}^{(1)}, \ldots, \mathbf{Z}^{(S)} \mid \mathbf{Q}].$$

**Mutation Operators** A mutation operator based on CDFS may be useful in evolution programs for multisensor fusion:

$$\mathbf{p}_i'' = \begin{cases} \text{instantiated from a fresh CDFS} & \text{if } \mathcal{U}(0,1) < \text{MU}, \\ \mathbf{p}_i'' & \text{otherwise.} \end{cases}$$

where MU is the CDFS mutation probability. Thus, we *replace* a fraction, MU, of the offspring population by new individuals drawn from the subspace of CDFS parameters.

**Population Initialization** The choice of the initial population is important, since it determines the trajectory followed in the search space. Typically, each individual in the population is initialized randomly, so that the initial population contains a "good" sampling of the search space. In some cases, a CDFS based initialization may be useful. In this scheme, each individual in the initial population is drawn from the subspace of CDFS parameters.

A combination of the random and CDFS based initialization may be used, where only a fraction of the population is initialized using the CDFS.

### 8.5.3.2 *Differential Evolution Program for Multisensor Fusion*

In this section we present a specific differential evolution program, which will be used to solve the multisensor object recognition problems discussed in Part 3 of this book. In this algorithm, finding the minimal representation size interpretation is regarded as a search problem in the space of environment model parameters:

$$\mathbf{p} \overset{\Delta}{=} < \Xi(\boldsymbol{\theta}) > .$$

The search *sequentially* iterates through the structures in the environment model library, and computes the data correspondences for each instantiated environment model as described in Section 8.3.

**Representation** Given a preselected environment model structure, each individual represents its environment model parameters:

$$S(\mathbf{p}) = \boldsymbol{\theta}.$$

All points in the parameter space $\Theta$ must be representable by the specific form chosen for the environment model parameters.

**Fitness Evaluation** The fitness function is the total multisensor representation size (Equation 7.10):

$$\Phi(S(\mathbf{p})) = \mathcal{L}[\mathbf{q}, \mathbf{Z}^{(1)}, \dots, \mathbf{Z}^{(S)} \mid \mathbf{Q}].$$

For an individual, $S(\mathbf{p}) = \boldsymbol{\theta}$, the fitness is evaluated as follows:

(1) If $\boldsymbol{\theta} \notin \Theta$, the total multisensor representation size is defined to be $\infty$.

(2) Otherwise, for $\boldsymbol{\theta} \in \Theta$:

  (a) For each sensor:

   i. Extract the model features.
   ii. Compute the minimal representation size correspondence, using the algorithms described in Section 8.3 on page 128.

  (b) Compute the total minimal representation size for this parameter vector (for the minimal representation size correspondences just computed).

**Reproduction** The standard DE crossover operator is used (Section 8.5.2).

**Selection** The standard DE selection operator is used (Section 8.5.2).

**Initialization** The parameter vectors are drawn *randomly* (Section 8.2.1) the space of legal environment model parameters $\boldsymbol{\theta} \in \Theta$. This ensures a diverse initial population.

**Termination Condition** A parameter is considered $\eta$-percent converged if at least $\eta$-percent of the population shares the same value (within some prespecified *parameter tolerance*) as the best individual in the population, for that parameter. The *population* is considered to be $\eta$-percent converged when all the parameters have $\eta$-percent converged.

The evolution is terminated when either **(a)** the population reaches a certain desired level of convergence, or **(b)** the maximum number of generations is exceeded, or **(c)** the maximum time limit is exceeded.

## 8.6 Choosing a Multisensor Fusion Algorithm

The minimal representation size interpretation may be found using either a hypothesize and test algorithm, described in Section 8.4, or the evolution program described in Section 8.5.3.2.

The advantages of using an evolution program include: **(a)** modularity, since incorporating additional sensors is straightforward, **(b)** it can fine-tune the parameters around the global minimum. On the other hand, there are no analytical results which guarantee finding the minimum or promise convergence.

The advantages of using a hypothesize and test algorithm include: **(a)** takes polynomial time, **(b)** guarantees correctness with a parameter resolution dependent on the observed data accuracy. On the other hand, incorporating additional sensors requires solving additional CDFS systems of equations.

A comparison of the hypothesize and test algorithm, and a differential evolution program is presented in Section 10.5.2. In practice, using hypothesize and test algorithm works well when the number of sensors, data features, model features, and observation errors are small, and the resulting CDFS are easily solved. Otherwise, an evolution program is preferable.

### 8.6.1 *Choosing an Evolution Program*

In traditional GAs, the representation, $S(\mathbf{p})$, is a *binary string*, based on the biological analogy with chromosomes which encode phenotypic information. Some analytical results have been established for this choice of representation, making it somewhat attractive to researchers. This binary string representation is easily applied to a wide variety of problem domains due to its simplicity; however, the use of problem specific representations may lead to more efficient evolution programs. Specifically, for parameter optimization problems, a *floating point string* representation has been found to be much more suitable in practice [136]. In parameter optimization problems, *fine local tuning* is necessary, to compute the problem parameters at a suitable resolution. The binary string representation has "inherent difficulties in performing local search for numerical applications" [136, page 107]. The length of binary strings increases with the desired parameter resolution, increasing the dimensionality of the representation space. Indeed, our experience with binary string GAs on multisensor object recognition problems (see Section 13.2.2) has confirmed these limitations for three-dimensional multisensor object recognition problems.

Historically, the ES and EP were developed specifically for solving parameter optimization problems, and used a floating point string representation. In these algorithms, mutation operators generate offspring by adding a zero mean *Gaussian* perturbation to the individuals in the current population. The problem representation $S(\mathbf{p})$ includes the floating point parameter vector, and also the *self-adaptation* parameters comprising of the standard deviations of the perturbations, used in the mutation operator. As the population converges, these standard deviations are reduced (also adapted by the evolutionary algorithm), thus accomplishing fine local tuning. However, several control parameters must be carefully tuned in practice, for successfully using these algorithms.

In DE also, the representation, $S(\mathbf{p})$, is a floating point string of parameter vectors. Here the crossover operator generates offspring by adding a differential perturbation the individuals in the current population. The magnitude of these differential perturbations remains scaled to a size appropriate for the population—as the population cluster shrinks, so do the differential perturbations, thus accomplishing fine local tuning.

In Sections 10.5.2 and 11.5.2, we discuss the performance of DE for two and three-dimensional multisensor object recognition problems, respectively. Our experience with DE on these problems has shown that it works quite well in practice. This has also been the experience of other researchers with the DE algorithm [156, 190]. In practice, the DE is also attractive due to its simple form which requires only a few control parameters.

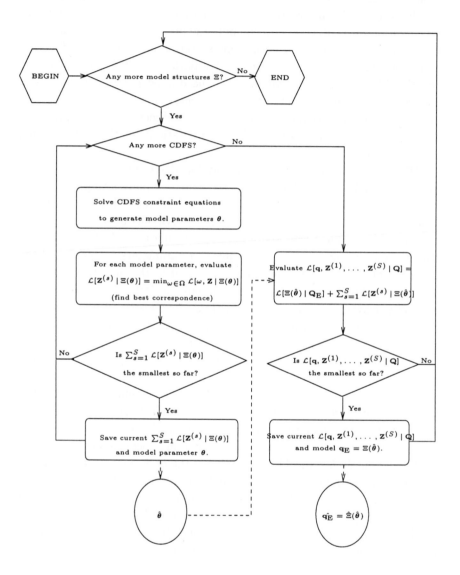

Fig. 8.3  A polynomial time hypothesize and test algorithm. Solid lines indicate control flow, while dotted lines indicate data flow.

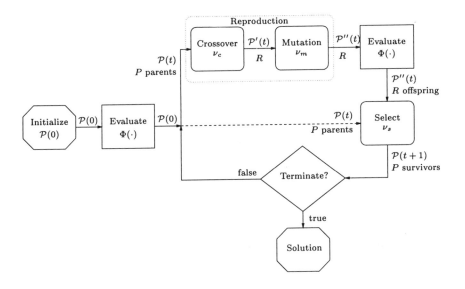

Fig. 8.4   An evolution program template.

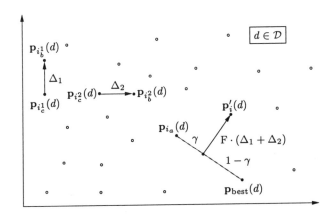

Fig. 8.5   DE crossover operator for $K = 2$ differentials. The differential perturbations $\Delta_1$ and $\Delta_2$ are applied in a randomly chosen projection, $d \in \mathcal{D}$, of the parameter space.

# PART 3

# Applications

## Chapter 9

# Applying the Abstract Framework to Concrete Problems

In this chapter we summarize the abstract framework, developed in Part 2, and outline a recipe for applying it to concrete problems.

## 9.1 Summary of the Abstract Framework

The multisensor fusion and model selection framework uses a *minimal representation size* criterion to choose among alternative models, number of parameters, parameter resolutions, and correspondences. The *representation size* or description length of an entity is defined as the length of the shortest length program that reconstructs the entity. The observed data, thought to be arising from a library of environment models, is encoded with respect to one of these models. The minimal representation size criterion selects the model which minimizes the total multisensor data representation size (Equation 7.10), and leads to a choice among alternative models which trades-off between the size of the model (e.g. number of parameters) and the representation size of the (encoded) residuals, or errors. Intuitively, the simplest most likely representation is selected as the preferred model for a given estimator. The model selection properties of the minimal representation size multisensor fusion framework are summarized below:

- **Data subsample selection (Correspondence model selection):** The framework trades-off unmodeled data feature representation size with encoded error residuals, to select the minimal representation size correspondence. The subsampled (modeled) data features are determined by the *sensor precision* (Section 7.4.3).

151

- **Environment model parameterization and data scaling:**
  The environment model parameter values are defined by the sensor
  constraints due to the subsampled (modeled) data features, which
  get "scaled" on the basis of *sensor accuracy,* and "weighted" on the
  basis of *sensor resolution* (Section 7.4.2).
- **Environment model selection:** The framework selects the environment model class, number of parameters, and a parameter
  resolution appropriate for the observed multisensor data, by trad-
  ing the environment model size against the sensor representation
  size, for the selected correspondences. The environment model *parameter resolution* becomes finer, as the sensor accuracy becomes
  finer, or as the number of data features increases (Section 7.4.1).

The model selection properties of this framework are complementary to the
estimation process itself, and the framework chooses an effective combination of model structure and parameter estimation method for a given class
of problems.

## 9.2    Recipe for Applying the Abstract Framework

To apply the abstract multisensor fusion and model selection framework to
a concrete problem, the following steps must typically be followed:

- Develop a model of the concrete problem domain:
  - (1) Specify an environment model library:
    - (a) Specify a representation for the various environment mod-
      el structures and parameters, which may be used to de-
      scribe the underlying physical environment.
    - (b) Develop expressions for environment model representation
      size (Section 7.3.1).
  - (2) Specify a sensor model for each sensor:
    - (a) Specify the sensor calibration.
    - (b) Specify the data conditioning needed to transform raw
      sensor data into a form suitable for use in the sensor con-
      straint equation (Figure 7.3).
    - (c) Specify the model feature extractor, which generates mod-
      el features (Figure 7.3) from an environment model rep-

resentation developed in Step 1a.

(d) Develop the sensor constraint equation, which relates data features to model features (Figure 6.1). Often these constraints can be obtained from physical modeling of the sensor.

(e) Specify an uncertainty model, which characterizes the observation errors and the sensor measurement space. May be specified in terms of either:

    i. Sensor accuracy and precision (Section 6.3.1), or

    ii. Probability distributions (Section 6.3.2).

The more natural uncertainty description should be preferred.

(f) Develop expressions for both modeled and unmodeled data feature representation size, using this uncertainty model with the constraint equation developed in Step 2d (Section 7.3.2.1).

(g) Specify a correspondence model, which may be either (Section 6.2):

    i. Many-to-One, or

    ii. One-to-One, or

    iii. Known.

The more natural natural description should be preferred (Section 7.3.3).

(h) Specify a correspondence model encoding algorithm:

    i. Fixed-length, for small data sets (Section 7.3.2.2).

    ii. Asymptotically optimal entropy, for large data sets (Section 7.3.2.2).

(i) Specify the sensor (correspondence + observed data) encoding scheme:

    i. Two-part (Section 7.3.2.2).

    ii. Mixture (Section 7.3.2.3).

Two-part encoding should be used when a discrete correspondence model must be explicitly selected; mixture encoding may be preferred for large data sets (see Section 7.3.3).

- Develop a multisensor fusion search algorithm to minimize the total multisensor representation size (Figure 7.3 and Equation 7.10). May use one of the generic algorithms described in Chapter 8:

(1) Hypothesize and Test (Section 8.4):

    (a) Develop CDFS systems of equations and their solutions (Section 8.2.2).

(2) Evolution programs (Section 8.5):

    (a) Develop a suitable concrete problem representation. A representation of the environment models (structure and parameters) suitable for evolutionary search must be chosen.

    A differential evolution program (Section 8.5.2) may be used, and is especially suited for continuous parameter optimization problems.

(3) Alternatively, another parameter estimation algorithm may be used, where the total representation size (Equation 7.10) serves only as a criterion for model selection (to choose among correspondence models, and between the various environment model structures and number of parameters).

These algorithms may be further optimized for a concrete problem domain, by using domain heuristics and exploiting specific domain knowledge.

The **MO**del-based **M**ultisensor **F**usion and **I**nterpretation **S**ystem (MOMFIS) object-oriented application framework described in Appendix B, captures this decomposition through its abstract classes. The specialization of these abstract classes, instantiates the abstract framework for various concrete problem domains.

# Chapter 10

# Multisensor Object Recognition in Two Dimensions

In this chapter, we illustrate the minimal representation size multisensor framework by applying it to the problem of two-dimensional object recognition and pose estimation, using touch and vision sensors. As shown in Figure 10.1, a fixed camera observes two-dimensional vertex features of an object, while tactile probes measure contact points on the object edges. The object identity and pose must be estimated by fusing the vision and tactile data.

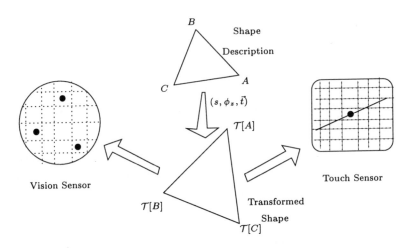

Fig. 10.1  A vision sensor observes the object vertices, while tactile sensors provide contact points on the object edges. The vision and tactile sensor data is fused to determine two-dimensional pose and the identity of the object under observation.

The objects are two-dimensional polygons parameterized by four pose parameters (scale, orientation, and translation), and are discussed in Section 10.1. The vision and the tactile sensor models are introduced in Section 10.2. In Section 10.3, we discuss the CDFS resulting from tactile visual data, and in Section 10.4, summarize the multisensor object recognition and pose estimation algorithm. In Section 10.5, we present simulation results, which serve to illustrate the minimal representation size multisensor fusion framework.

The notation used to describe the environment and sensor models is summarized in Table 10.1.

## 10.1   Environment Models

An environment model, $\mathbf{q}_E$, is specified by a polygonal shape structure, $\Xi$, associated with four pose parameters $\boldsymbol{\theta} = (s, \phi_z, \vec{t})$, where $s$ is the scale factor, $\phi_z$ is the rotation angle (counterclockwise), and $\vec{t}$ is the translation in the two-dimensional plane; thus $\mathbf{q}_E = \Xi(s, \phi_z, \vec{t})$. The environment model library $\mathbf{Q}_E$ is an enumeration of the shape structures under consideration, in all possible poses. Therefore, the model size is given by the number of bits needed to specify the shape structure, and the number of bits to specify its pose:

$$\mathcal{L}[\mathbf{q}_E \mid \mathbf{Q}_E] = \log_2(L+1) + \sum_{d=1}^{D} c_{\theta_d} = c_{\mathbf{q}_E}$$

where $L$ is the number of shape structures, and $D$ is the number (four) of pose parameters. The model size is a constant for all shape models since they all have the same number of pose parameters. Thus, the model selection problem is essentially the object identification problem.

Let the polygonal shape structure, $\Xi$, be described by a set of *vertices* $\{\vec{p}\}$, and a set of *edges* $\{e\}$ in some local object-centric coordinate frame. An edge is a directed line segment joining vertex $\vec{p}^a$ to vertex $\vec{p}^b$, and denoted $e = <\vec{p}^a, \vec{p}^b>$.

The pose parameters $\boldsymbol{\theta} = (s, \phi_z, \vec{t})$ can be regarded as a description of the local object coordinate frame in the world coordinate frame [41]. An environment "shape" model is expressed as $\mathbf{q}_E = \Xi(s, \phi_z, \vec{t}) \equiv \mathcal{T}[\Xi]$ where $\mathcal{T}[\cdot]$ denotes the pose transformation needed to describe the shape in the world coordinates. Thus, the vertices and edge of the shape model are obtained by applying the pose transformation to the shape structure, and are denoted by $\vec{p}' = \mathcal{T}[\vec{p}]$ and $\mathbf{e}' = \mathcal{T}[\mathbf{e}]$ respectively.

Using *complex numbers* to represent a point in the two-dimensional plane, we get:

$$\vec{p}' = \mathcal{T}[\vec{p}] = se^{i\phi_z}\vec{p} + \vec{t}$$
$$\mathbf{e}' = \mathcal{T}[\mathbf{e}] = < \mathcal{T}[\vec{p}^a], \mathcal{T}[\vec{p}^b] >$$

where the primes (′) indicate the transformed values.

### 10.1.1  *Pose Representation for Evolution Programs*

The two-dimensional pose is expressed as a four parameter vector,

$$\theta = (s, \phi_z, t_x, t_y),$$

where $s \in [s^{\min}, s^{\max}]$, $\phi_z \in [-\pi, \pi)$, $t_x \in [t_x^{\min}, t_x^{\max}]$, and $t_y \in [t_y^{\min}, t_y^{\max}]$.

## 10.2  Sensor Models

### 10.2.1  *Vision*

#### 10.2.1.1  *Calibration*

We assume that the camera image plane is parallel to the two-dimensional world under observation. The two-dimensional camera calibration consists of a transformation which maps the two-dimensional pixel coordinates to the two-dimensional world coordinates.

#### 10.2.1.2  *Data Features*

The vision sensor provides object vertex features, which may be obtained by applying low-level vision processing to the images captured from a camera.

The raw vision data features in pixel coordinates are converted or *conditioned* into the world coordinates, by applying the two-dimensional camera calibration:

$$\mathbf{z}^{(v)} \equiv \vec{r}.$$

The set of such image vertex features $\{\vec{r}\}$ comprise the vision data features.

### 10.2.1.3 *Model Features*

The shape model vertices comprise the set of vision model features:

$$\mathbf{y}^{(v)} \equiv \vec{p}\,'.$$

### 10.2.1.4 *Constraint Equation*

Ideally, an observed vision data feature must coincide with a shape vertex feature:

$$\vec{r} = \vec{p}\,' = se^{\imath \phi_z} \vec{p} + \vec{t}. \tag{10.1}$$

We assume that at most one vision image vertex can correspond to a shape model vertex, i.e., the vision correspondence must be *one-to-one* (Section 6.2.2).

### 10.2.1.5 *Uncertainty Model*

The measurement space of the vision data features is the finite image plane, which is subdivided into $K_U \times K_V$ rectangular pixels, each of size $\sigma_u \times \sigma_v$ in the world coordinate frame.

The sensor accuracy is given by approximating a rectangular pixel by an elliptical uncertainty region (Section 6.3.1.1, Figure 6.9), with the semi-principal axis lengths proportional to $\sigma_u$ and $\sigma_v$:

$$A = \begin{bmatrix} 1/\sigma_u^2 & 0 \\ 0 & 1/\sigma_v^2 \end{bmatrix}$$

and radius $r = 1$.

The sensor precision is given the number of bits needed to address the measurement space (Section 6.3.1):

$$\lambda = \log_2(K_U \cdot K_V + 1).$$

### 10.2.1.6 *Representation Size*

The ideal vision data feature which can arise from a shape model vertex feature is the shape model vertex itself:

$$\tilde{\vec{r}} = \vec{p}\,'$$

and comprises the DCM (Section 7.3.2.1). Since the DCM is a singular point, $\mathcal{L}[\tilde{\vec{r}} \,|\, \vec{p}\,'] = 0$.

The error residual representation size (Section 7.3.2.1) is calculated in the world frame, using the vision sensor accuracy: (Section 6.3.1):

$$\mathcal{L}[\vec{r} - \tilde{\vec{r}} \,|\, \tilde{\vec{r}}] \;=\; \mathcal{I}\!\left(\left[(\vec{r} - \vec{p}\,')^{\mathrm{T}} \frac{A}{r^2} (\vec{r} - \vec{p}\,')\right]\right)$$

where $n = 2$ in Equation (6.6).

The modeled data feature representation size is given by (Equation 7.15):

$$\begin{aligned}
\mathcal{L}[\vec{r} \,|\, \mathbf{y} = \vec{p}\,'] &= \mathcal{L}[\tilde{\vec{r}} \,|\, \vec{p}\,'] + \mathcal{L}[\vec{r} - \tilde{\vec{r}} \,|\, \tilde{\vec{r}}] \\
&= \mathcal{I}\!\left(\left[(\vec{r} - \vec{p}\,')^{\mathrm{T}} \frac{A}{r^2} (\vec{r} - \vec{p}\,')\right]\right)
\end{aligned} \tag{10.2}$$

The unmodeled data feature representation size (Section 7.3.2.1) is given by the vision sensor precision:

$$\mathcal{L}[\vec{r} \,|\, \mathbf{y_0}] = \lambda = \log_2(K_U \cdot K_V + 1). \tag{10.3}$$

### 10.2.2   *Touch*

#### 10.2.2.1   *Data Features*

Tactile sensors provide two-dimensional contact points, expressed in world coordinates:

$$\mathbf{z}^{(\mathrm{v})} \equiv \vec{w}.$$

The set of such contact points $\{\vec{w}\}$ comprises the touch data features.

#### 10.2.2.2   *Model Features*

The shape model *edges*, $\{\mathbf{e}'\}$, comprise the touch model features:

$$\mathbf{y}^{(\mathrm{t})} \equiv \mathbf{e}'.$$

We assume that contact is made only on the object edges.

### 10.2.2.3   *Constraint Equation*

Since the tactile sensors contact the object on its edges, an observed contact point data feature must lie on an object edge:

$$\exists \zeta \in [0,1], \quad \vec{w} = \zeta \cdot \mathcal{T}[\vec{p}^{\,b}] + (1-\zeta) \cdot \mathcal{T}[\vec{p}^{\,a}]$$
$$= s e^{i\phi_z} [\zeta \vec{p}^{\,b} + (1-\zeta)\vec{p}^{\,a}] + \vec{t} \qquad (10.4)$$

where the contact point $\vec{w}$ lies on the finite object edge $e' = \;<\; \mathcal{T}[\vec{p}^{\,a}], \mathcal{T}[\vec{p}^{\,b}] \;>$, and the parametric coordinate $\zeta$ defines the location of the contact point on the object edge.

Several contacts may be made on an object edge, resulting in a *many-to-one* correspondence.

### 10.2.2.4   *Uncertainty Model*

The contact point observation errors are described by an elliptical uncertainty region (Figure 6.9), whose semi-principal axis lengths are given by $\sigma_{x_t}$, $\sigma_{y_t}$ in the world coordinate frame:

$$A = \begin{bmatrix} 1/\sigma_{x_t}^2 & 0 \\ 0 & 1/\sigma_{y_t}^2 \end{bmatrix}$$

and whose radius $r = 1$.

The measurement space of contact points is the finite two-dimensional region of points reachable by the tactile sensors. Let us assume that it is a larger ellipse with the same shape as the uncertainty region ellipse, but with a radius $R$. The touch sensor precision is given by:

$$\lambda = \log_2 \left( \left[ \frac{R}{r} \right]^2 + 1 \right),$$

where $n = 2$ in Equation (6.7).

### 10.2.2.5   *Representation Size*

The DCM is the set of all points lying on the finite object edge:

$$\text{DCM}(e') = \{ \tilde{\vec{w}} : \tilde{\vec{w}} \in e' \}.$$

The DCM is one-dimensional, $\tilde{n} = 1$, and its representation size is given by, $\mathcal{L}[\tilde{\vec{w}} \,|\, e'] = \tilde{c}_\zeta = \tilde{c}_t$ (Equation 7.16), where $\tilde{c}_\zeta$ is the fixed number of bits used to encode the parametric coordinate $\zeta \in [0,1]$, in the constraint Equation (10.4).

Given an observed touch data feature, $\vec{w}$, the "ideal" touch data feature, $\tilde{w}$, is the point *closest* to the observed data feature, lying on the edge $e'$:

$$\tilde{w} = \tilde{\zeta} \cdot \mathcal{T}[\vec{p}^b] + (1 - \tilde{\zeta}) \cdot \mathcal{T}[\vec{p}^a]$$

where $\tilde{\zeta}$ is a parametric coordinate along the edge $e' = \, <\mathcal{T}[\vec{p}^a], \mathcal{T}[\vec{p}^b]>$, computed as follows. From the definition of $\tilde{w}$, we have $(\vec{w} - \tilde{w}) \perp (\mathcal{T}[\vec{p}^b] - \mathcal{T}[\vec{p}^a])$. Therefore[*]:

$$(\vec{w} - \tilde{w})^* (\mathcal{T}[\vec{p}^b] - \mathcal{T}[\vec{p}^a]) + (\vec{w} - \tilde{w})(\mathcal{T}[\vec{p}^b] - \mathcal{T}[\vec{p}^a])^* = 0,$$

which can be solved to find $\tilde{\zeta}$. Since the DCM point must be contained in the finite edge, the value of $\tilde{\zeta}$, thus found, is *clipped* to lie in $[0,1]$:

$$\tilde{\zeta} = \max(0, \min(1, \frac{\mathrm{Re}[(\vec{w} - \mathcal{T}[\vec{p}^a])^*(\mathcal{T}[\vec{p}^b] - \mathcal{T}[\vec{p}^a])]}{(\mathcal{T}[\vec{p}^b] - \mathcal{T}[\vec{p}^a])^*(\mathcal{T}[\vec{p}^b] - \mathcal{T}[\vec{p}^a])})).$$

The error residual representation size is given by:

$$\mathcal{L}[\vec{w} - \tilde{w} \,|\, \tilde{w}] = \mathcal{I}(\left[(\vec{w} - \tilde{w})^{\mathrm{T}} \frac{A}{r^2}(\vec{w} - \tilde{w})\right]).$$

where $n = 2$, in Equation (6.6).

The modeled data feature representation size is given by (Equation 7.15):

$$\begin{aligned}
\mathcal{L}[\vec{w} \,|\, \mathbf{y} = e'] &= \mathcal{L}[\tilde{w} \,|\, e'] + \mathcal{L}[\vec{w} - \tilde{w} \,|\, \tilde{w}] \\
&= \tilde{c}_t + \mathcal{I}(\left[(\vec{w} - \tilde{w})^{\mathrm{T}} \frac{A}{r^2}(\vec{w} - \tilde{w})\right]).
\end{aligned} \tag{10.5}$$

The unmodeled data feature representation size (Section 7.3.2.1) is given by the touch sensor precision:

$$\mathcal{L}[\vec{w} \,|\, \mathbf{y}_0] = \lambda = \log_2(\left[\frac{R}{r}\right]^2 + 1). \tag{10.6}$$

---

[*]Given two complex numbers, $z_1$ and $z_2$, they are orthogonal iff $\mathrm{Re}[z_1 z_2^*] = 0$, where $^*$ denotes the conjugate. Since $\mathrm{Re}[z] = (z + z^*)/2$, therefore, $z_1 \perp z_2$ iff $z_1 z_2^* + z_1^* z_2 = 0$.

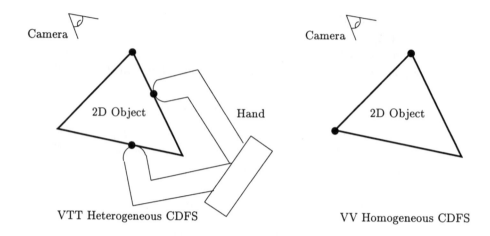

Fig. 10.2   VV and VTT constraining data feature sets.

## 10.3   Constraining Data Feature Sets

The scale, position and orientation (a total of four parameters) of a given polygonal shape are hypothesized using image vertex features from the vision sensor and contact point features from the touch sensor.

The vision (V) constraint equation of Equation (10.1) gives us 2 (real) equations in 4 unknowns (the pose parameters); the touch (T) constraint equation of Equation (10.4) gives gives us 2 (real) equations in 5 unknowns (the 4 pose parameters, and $\zeta$). Accordingly the CDFS combinations are: VV using two vision constraints, VTT using one vision and two touch constraints, and TTTT using four touch constraints; some of these are illustrated in Figure 10.2.

Each CDFS system of equations results in a system of *linear equations*, which can be solved exactly and may result in multiple pose solutions, as described in Appendix F. The total number of CDFS system of equations is given by:

$$\binom{N^{(v)}}{2}\binom{M^{(v)}}{2}2! + \binom{N^{(v)}}{1}\binom{M^{(v)}}{1}\binom{N^{(t)}}{2}\binom{M^{(t)}}{2}2! + \binom{N^{(t)}}{4}\binom{M^{(t)}}{4}4!$$

where $N^{(v)}$ and $M^{(v)}$ are the number of vision data and model features respectively, while $N^{(t)}$ and $M^{(t)}$ are the number of touch data and model features respectively.

## 10.4   Multisensor Object Recognition Algorithms

The polynomial time hypothesize and test algorithm, described in Section 8.4, can be used to find the minimal representation size interpretation from vision and touch data. A *two-part* encoding scheme is used for both sensors and the data correspondences are computed as described in Section 8.3.1. The total time complexity is given by:

$$O(M^4 \cdot N^4) \times [O(M \cdot N) + O(M \cdot N \cdot \min(M, N)) + O(M \cdot N) + O(M \cdot N)]$$
$$= O(M^5 \cdot N^5 \cdot \min(M, N)).$$

The differential evolution (DE) program, described in Section 8.5.3.2 on page 142 may also be used to search for the minimal representation size pose and correspondences. The pose parameter representation is described in Section 10.1.1.

## 10.5   Computer Simulations

The two-dimensional object shapes used in the computer simulations, are shown in Figure 10.3. To generate a simulated environment model, an object shape is chosen randomly from the environment model library, and its parameters are instantiated randomly, $s \in [0.5, 1.5]$, $\phi_z \in [-\pi, \pi)$, $t_x \in [-50, -50]$, $t_y \in [-50, -50]$.

The vision and the tactile sensors were modeled by a two-dimensional ellipsoidal uncertainty region of radius 1 (Figure 6.9 on page 85), and a measurement space of radius 100. These uncertainty models were used to randomly generate the multisensor data from the randomly generated environment model.

A one-to-one correspondence model was used for the vision sensor, while a many-to-one correspondence model was used for the tactile sensor. In both cases, a two-part fixed-length encoding algorithm (Section 7.3.2.2) was used to compute the representation size, using the sensor models developed in Sections 10.2.1 and 10.2.2.

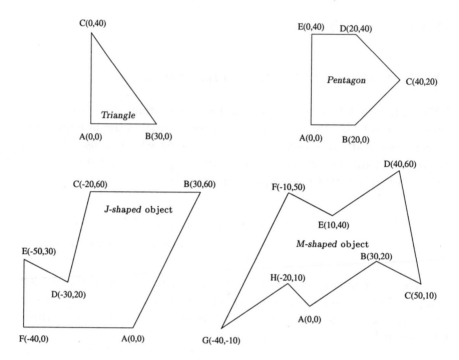

Fig. 10.3   Two-dimensional object shapes used in computer simulations.

### 10.5.1   *Model Selection Performance*

The model selection performance of the minimal representation size multisensor technique was evaluated on 100 randomly generated problems. A typical problem size was chosen, consisting of 20 vision data features of which 50% were spurious, and 10 tactile data features of which 20% were spurious.

A 40 member DE/best/2 differential evolution algorithm with $F = 0.5$, and $CR = 0.3$ (Sections 8.5.2 and 8.5.3.2), was used to search for the minimal representation size pose and the correspondences for each sensor. The search is terminated when the population reaches 50% convergence (parameter tolerance of 0.001), or when the maximum time limit of 300 CPU seconds is exceeded, or when a maximum of 1000 generations have been completed. The search is considered to have correctly converged (to the global minimum), if it reaches within 1 bit of the simulated multisensor

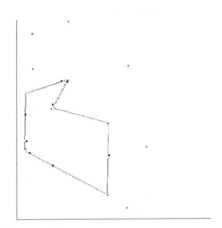

Fig. 10.4   Simulation results for the *J-shaped* object, showing the reference pose super-imposed on the estimated pose found by fusing vision and touch data. The vision data features are shown as gray dots, while the touch data features are shown as black dots.

data representation size.

The model selection performance of the technique is assessed from the number of misclassifications, and the average pose and correspondence errors among those correctly classified. The pose error is given by the scale error: $\hat{s}/s - 1$, the orientation error: $\hat{\phi}_z - \phi_z \mod 2\pi$, and the translation error: $|\hat{\vec{t}} - \vec{t}|$, where $\hat{x}$ denotes the estimated value of $x$. The correspondence error is given by the number of mis-assignments in the estimated correspondence.

A typical simulation sample is shown in Figure 10.4. The simulation results on 100 randomly generated problems for two different environment model libraries are tabulated in Table 10.2 on page 171. The mean errors and standard deviations are tabulated among those DE searches which correctly converged, and the correct object shape was chosen. There was one misclassification when using the environment model library with four object shapes. For that particular randomly generated configuration of data features, the multisensor data from a *J-shaped* object was better explained by the simpler *Pentagon* object.

### 10.5.2 *Search Performance*

The time complexity of the differential evolution (DE) and the hypothesize and test (H&T) search algorithms was evaluated on different problem sizes. The environment model library had exactly one environment model: the *J-shaped* object. The search algorithms were run on randomly generated problems with varying number of vision and touch data features. A 40 member DE/best/2 differential evolution algorithm with F = 0.5, and CR = 0.3 was used. The DE search is terminated when the population reaches 50% convergence (parameter tolerance of 0.001), or when the maximum time limit of 600 CPU seconds is exceeded, or when a maximum of 1000 generations have been completed. Results are plotted for the DE runs which correctly converged (reached within 1 bit of the simulated multisensor data representation size). The H&T algorithm is run on these same problems, and its results are plotted on the same graph.

Figure 10.5 shows the CPU time against the number of vision data features, for a fixed number of touch data features. Figure 10.6 shows the

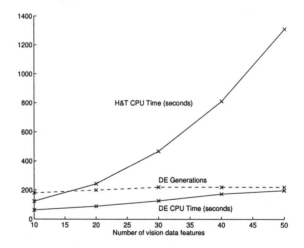

Fig. 10.5 CPU time (SPARC 20) vs. the number of vision data features for the differential evolution (DE) and the hypothesize and test (H&T) search algorithms. The vision sensor had 50% outliers, while the touch sensor produced 10 data features with 50% outliers.

number of minimal representation size correspondence evaluations (Section 8.3) against the number of vision data features. Figure 10.7 shows the

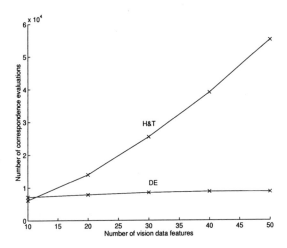

Fig. 10.6 Number of minimal representation size correspondence evaluations vs. the number of vision data features for the differential evolution (DE) and the hypothesize and test (H&T) search algorithms.

CPU time against the number of touch data features, for a fixed number of vision data features. Figure 10.8 shows the number of minimal representation size correspondence evaluations against the number of touch data features.

From these plots, we see that the DE converges in approximately 200 generations, and requires approximately 8000 minimal representation size correspondence evaluations. These values seem to change very little with the problem size. The CPU time increases linearly with the number of data features $N$. This also follows from the theoretical analysis (see Section 8.3.1), since the time taken to compute a one-to-one vision minimal representation size correspondence grows as $O(M \cdot N \min(M, N))$ which is linear in $N$ when $M \leq N$ (as is the case in these simulations), while the time taken to compute a many-to-one touch correspondence grows as $O(M \cdot N)$ which is linear in $N$. In fact, as the number of data features increases, the DE appears to converge correctly more consistently. This may be due to the fact that the global minimum is deeper as the number of data features increases. As seen from Table 10.2, nearly 75% of the DE searches correctly converged on 100 randomly generated problem instances. This number may be improved further, by tuning the DE search parameters $P$,

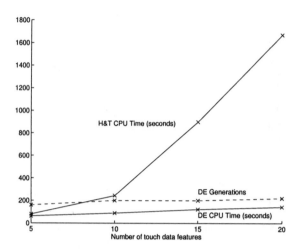

Fig. 10.7   CPU time (SPARC 20) vs. the number of touch data features for the differential evolution (DE) and the hypothesize and test (H&T) search algorithms. The touch sensor had 50% outliers, while the vision sensor produced 20 data features with 50% outliers.

F, and CR.

For the H&T algorithm, the number of minimal representation size correspondence evaluations grows quadratically with the number of vision data features, and as a fourth order polynomial with the number of touch data features, as predicted by the theoretical analysis (Section 10.3). Therefore, the CPU time grows as a cubic with the number of vision data features, and as a fifth-order polynomial with the number of touch data features (since the time to compute each minimal representation size correspondence grows linearly with the number of data features).

For small problem sizes ($N, M \approx 10$), both DE and H&T algorithm exhibit similar performance, and are quite practical. However, as the problem size increases, the H&T algorithm quickly becomes impractical and a DE is preferred despite the lack of any guarantee of finding the correct solution. In practice, several DE searches may be executed in (in parallel), to improve the reliability of the search.

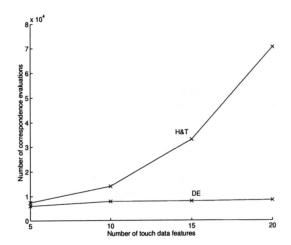

Fig. 10.8 Number of minimal representation size correspondence evaluations vs. the number of touch data features for the differential evolution (DE) and the hypothesize and test (H&T) search algorithms.

Table 10.1   Environment and sensor models in two-dimensional multisensor object recognition.

| Symbol | Meaning |
|---|---|
| | *Environment Model* |
| $\Xi$ | polygonal shape description: geometry described in local object coordinates |
| $\theta$ | pose $\mathcal{T}[\cdot]$: scale $s$, rotation $\phi_z$, and translation $\vec{t}$ |
| $\mathbf{q_E}$ | polygonal shape model: $\mathbf{q_E} = \Xi(s, \phi_z, \vec{t}) \equiv \mathcal{T}[\Xi]$ |
| $\mathbf{Q_E}$ | shape library: set of objects under consideration in all possible poses |
| | *Vision Sensor (V)* |
| $\Psi$ | vision accuracy |
| $\Upsilon$ | measurement space is the image size |
| $\alpha$ | two-dimensional camera calibration |
| $\mathbf{z}$ | image vertex: $\vec{r}$ |
| $\mathbf{y}$ | shape vertex: $\vec{p}\,'$ |
| $\omega$ | one-to-one correspondence between image vertices and shape model vertices |
| $\mathbf{h}$ | a vision data vertex must coincide with an object vertex: $\vec{r} = \vec{p}\,'$ |
| $\mathcal{F}$ | object vertex extraction from polygonal shape: $\{\vec{p}\,'\} = \mathcal{F}(\Xi(s, \phi_z, \vec{t}); )$ |
| | *Touch Sensor (T)* |
| $\Psi$ | touch accuracy |
| $\Upsilon$ | measurement space is the finite region accessible by the tactile sensor |
| $\alpha$ | tactile sensor calibration |
| $\mathbf{z}$ | contact point: $\vec{w}$ |
| $\mathbf{y}$ | shape edge: $\mathbf{e}'$ |
| $\omega$ | many-to-one correspondence between contact points and shape model edges |
| $\mathbf{h}$ | a contact point must lie on a shape model edge |
| $\mathcal{F}$ | object edge extraction from polygonal shape: $\{\mathbf{e}'\} = \mathcal{F}(\Xi(s, \phi_z, \vec{t}); )$ |
| $\zeta$ | intermediate variable |

Table 10.2   Model selection performance for two environment model libraries, on 100 randomly generated problems.   Each problem had 20 vision data features with 50% outliers, and 10 touch data features with 20% outliers.

|  | Environment Model Library | |
|---|---|---|
|  | *J-shape* *M-shape* | *Triangle, Pentagon,* *J-shape, M-shape* |
| Trials | 100 | 100 |
| Correctly Converged | 73 | 75 |
| Misclassified | 0 | 1 |
| Scale error, mean | -0.000348 | 0.000177 |
| standard deviation | 0.007351 | 0.008681 |
| Orientation error, mean | 0.031419° | 0.014398° |
| standard deviation | 0.454961° | 0.620666° |
| Translation error, mean | 0.402514 | 0.441040 |
| standard deviation | 0.232942 | 0.227309 |
| Vision correspondence error, mean | 0.808219 | 0.608108 |
| standard deviation | 0.892212 | 0.841029 |
| Touch correspondence error, mean | 0.849315 | 0.756757 |
| standard deviation | 0.793572 | 0.790540 |
| Representation size deviation, mean | -4.979178 | -4.169054 |
| standard deviation | 3.498506 | 3.065178 |

## Chapter 11

# Multisensor Object Recognition in Three Dimensions

In this chapter, we apply the minimal representation size multisensor framework to the problem of three-dimensional object identification and pose estimation, using a variety of sensors. Let us consider the multisensor fusion problem, introduced in Section 1.2 on page 2 (Figure 1.1). A robot hand holds an object in front of a camera, and the information from the vision, tactile, and grasp sensors may be fused, to determine the identity and pose of the object in the scene, in some prespecified world coordinate frame.

The environment models are polyhedral objects, parameterized by the pose (position and orientation), and are discussed in Section 11.1. In Section 11.2, we introduce three sensor models: vision, touch, and grasp*. The *vision* sensor provides two-dimensional image vertex features, obtained by processing camera images. The *touch* sensor provides three-dimensional contact points, while the *grasp* sensor provides contact points and contact surface normals, obtained from the finger-tips of a robot hand grasping an object.

In Section 11.3, we discuss the problem of shape matching using only the grasp sensor, and present some simulation and experimental results. In Section 11.4, we discuss the problem of pose estimation using vision and touch sensors, and present some simulation results. Detailed laboratory experiments were conducted for this problem, and are described in Chapter 12.

The notation used to describe the environment model and the various

---

*The difference between the "grasp" and "touch" sensor is that, "grasp" provides both contact point and surface normal information, whereas "touch" provides only the contact point information.

173

sensor models is summarized in Table 11.1.

## 11.1   Environment Models

An environment model, $\mathbf{q}_E$, is specified by a polyhedral shape structure, $\Xi$, and six pose parameters $\boldsymbol{\theta} = (R, \vec{t}\,)$, where $R$ is the rotation and $\vec{t}$ is the translation; thus $\mathbf{q}_E = \Xi(R, \vec{t}\,)$. The environment model library $\mathbf{Q}_E$ is an enumeration of the shape structures under consideration, in all possible poses. Therefore, the model size is given by the number of bits needed to specify the shape structure, and the number of bits to specify its pose:

$$\mathcal{L}[\mathbf{q}_E \,|\, \mathbf{Q}_E] = \log_2(L+1) + \sum_{d=1}^{D} c_{\theta_d} = c_{\mathbf{q}_E}$$

where $L$ is the number of shape structures, and $D$ is the number (six) of pose parameters. The model size is a constant for all shape models since they all have the same number of pose parameters. Thus, the model selection problem is essentially the object identification problem.

Let the polyhedral shape structure, $\Xi$, be described by a set of *vertices* $\{\vec{p}\,\}$, and a set of *faces* $\{\mathbf{f}\}$ in some local object-centric coordinated frame. The faces are polygonal, and completely described by the list of object vertices on their boundary. A face may also be specified by an *outward* surface normal $\vec{m}$, and a perpendicular distance $c$ from the origin; thus $\mathbf{f} = (\vec{m}, c)$. Alternatively a face may also be specified by three non-collinear vertices lying on it, denoted $\mathbf{f} = \{\vec{p}^{\,a}, \vec{p}^{\,b}, \vec{p}^{\,c}\}$. A list of vertices on the boundary is still needed to specify its finite polygonal shape.

The pose parameters $\boldsymbol{\theta} = (R, \vec{t}\,)$ can be regarded as a description of the local object coordinate frame in the world coordinate frame [41]. An environment "shape" model can be expressed as $\mathbf{q}_E = \Xi(R, \vec{t}\,) \equiv \mathcal{T}[\Xi]$ where $\mathcal{T}[\cdot]$ denotes the pose transformation needed to describe the shape in the world coordinates. Thus, the vertices and faces of the shape model are obtained by applying the pose transformation to the shape structure, and are denoted by $\vec{p}\,' = \mathcal{T}[\vec{p}\,]$ and $\mathbf{f}' = \mathcal{T}[\mathbf{f}]$ respectively. Expressing the rotation as a *unit quaternion q* (Appendix H):

$$\vec{p}\,' = \mathcal{T}[\vec{p}\,] \quad = \quad q\vec{p}\,q^* + \vec{t} \tag{11.1}$$

$$\vec{m}' = \mathcal{T}[\vec{m}] \quad = \quad q\vec{m}q^* \tag{11.2}$$

$$c' = \mathcal{T}[c] \quad = \quad c + \vec{t} \cdot (q\vec{m}q^*) \tag{11.3}$$

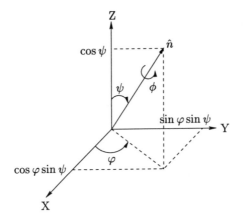

Fig. 11.1   A rotation of angle $\phi$ around the axis $\hat{n}$.

where the primes (') indicate the transformed values. A brief review of quaternions is presented in Appendix H.

### 11.1.1   *Pose Representation for Evolution Programs*

A rotation of angle $\phi \in [0, 2\pi)$, about an axis $\hat{n}$ is given by the unit quaternion (Appendix H),

$$q = e^{\hat{n}\frac{\phi}{2}} = \cos\frac{\phi}{2} + \hat{n}\sin\frac{\phi}{2}.$$

As shown in Figure 11.1, the axis of rotation, $\hat{n}$, can be represented by the spherical coordinates $(\varphi, \psi)$:

$$\hat{n} = (\cos\varphi\sin\psi,\ \sin\varphi\sin\psi,\ \cos\psi)$$

which lie on the *unit sphere*.

This representation is redundant, since a rotation of angle $\phi$ about the axis $\hat{n}$ is equivalent to a rotation of $-\phi$, about the axis $-\hat{n}$. Also, $q$ has a periodicity of $4\pi$ in $\phi$, and represents the same rotation as $-q$. To make the representation non-redundant, we restrict $\hat{n}$ to lie in the *upper hemisphere* (Figure 11.1):

$$\varphi \in [-\pi, \pi) \qquad \psi \in [0, \pi/2].$$

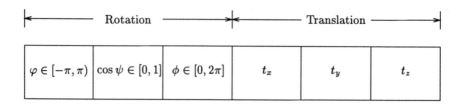

Fig. 11.2  Pose representation.

When, $\psi = \pi/2$, we restrict $\hat{n}$ to lie in the first quadrant:

$$\varphi \in [0, \pi/2] \quad \text{when} \quad \psi = \pi/2.$$

These bounds on the rotation parameters $\varphi, \psi, \phi$ guarantee a *unique* representation for every rotation.

A uniform sampling of the rotation parameter space $(\varphi, \psi, \phi)$, results in a non-uniform distribution of rotations—rotation axes, $\hat{n}$, are "bunched" up near the north pole. This can also be seen from the expression of the surface area of the unit sphere, as a function of the polar coordinates $(\varphi, \psi)$:

$$A(\varphi, \psi) = \int_0^\varphi \int_0^\psi \sin\psi \; d\psi \, d\varphi = (1 - \cos\psi)\varphi.$$

When $\varphi$ and $\psi$ are sampled uniformly, the area $A(\varphi, \psi)$ is sampled non-uniformly. However, choosing the pose parameters to be $(\varphi, \cos\psi, \phi)$, where $\cos\psi \in [0, 1]$, results in a uniform sampling of $A(\varphi, \psi)$ (when this new parameter space is sampled uniformly). In this parameter space, the distance between two representations is indicative of the "distance" in problem space. Also, an initial population is easily generated by uniformly sampling the parameter space, $(\varphi, \cos\psi, \phi)$.

The *translation* parameters are given by a vector,

$$\vec{t} = (t_x, t_y, t_z)$$

where each of the parameters is assumed to lie in a finite range, i.e. $t_x \in [t_x^{\min}, t_x^{\max}]$, $t_y \in [t_y^{\min}, t_y^{\max}]$, $t_z \in [t_z^{\min}, t_z^{\max}]$.

This pose representation is expressed as a six parameter vector,

$$\boldsymbol{\theta} = (\varphi, \cos\psi, \phi, t_x, t_y, t_z),$$

as shown in Figure 11.2.

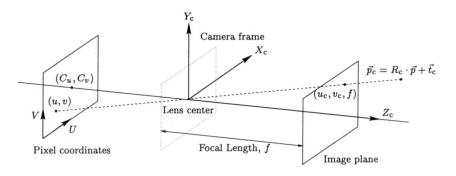

Fig. 11.3 The pin-hole camera model.

## 11.2 Sensor Models

### 11.2.1 *Vision*

#### 11.2.1.1 *Camera Calibration*

The *extrinsic* camera parameters [149] describe the transformation of a point expressed in the world coordinates, to a *lens centered* camera coordinate frame, whose $Z_c$-axis is coincident with the optical axis, as shown in Figure 11.3. The extrinsic camera parameters are described by the position, $\vec{t_c}$, and the orientation, $R_c$, of the world coordinate frame in the camera frame; thus

$$\vec{p_c} = R_c \cdot \vec{p} + \vec{t_c} \tag{11.4}$$

where $\vec{p_c}$ and $\vec{p}$ are the camera and world coordinates of the same point, respectively. The extrinsic parameters change when the camera is moved.

The *intrinsic* parameters [149] describe the transformation of a three-dimensional point expressed in the camera frame, to the two-dimensional pixel coordinates obtained from a digital image. They do not change when a camera is moved. We assume a *pin-hole* camera model, with an image plane located at a distance equal to the focal length, $f$, in front of the camera (Figure 11.3). The image plane coordinates of a point $\vec{p_c} = (x_c, y_c, z_c)$, are given by $(u_c, v_c)$ in the camera frame:

$$u_c = \frac{f}{z_c} \cdot x_c, \qquad v_c = \frac{f}{z_c} \cdot y_c. \qquad (11.5)$$

These image plane coordinates undergo a radial lens transformation, to give the *distorted* image plane coordinates, $(u_{cd}, v_{cd})$, in the camera frame:

$$u_{cd}\{\kappa \cdot (u_{cd}^2 + v_{cd}^2) + 1\} = u_c, \qquad v_{cd}\{\kappa \cdot (u_{cd}^2 + v_{cd}^2) + 1\} = v_c,$$

where $\kappa$ is the radial lens distortion coefficient. The distorted image plane coordinates are *reflected* about the lens center, and *sampled* by the frame-grabber to give the pixel coordinates $(u, v)$ in the digital image (Figure 11.3):

$$u = C_u + s_{\text{rate}} \cdot \frac{K_U}{K_{\text{ccd}}} \cdot \frac{(-u_{cd})}{\sigma_u}, \qquad v = C_v + \frac{(-v_{cd})}{\sigma_v},$$

where $(C_u, C_v)$ are the pixel coordinates of the lens center, $\sigma_u$ is the center-to-center distance between adjacent sensor elements in the $U$ (scan-line) direction, $\sigma_v$ is the center-to-center distance between adjacent elements in the $V$ direction, $K_{\text{ccd}}$ is the number of CCD sensor elements in the $U$ direction, $K_U$ is the number of pixels in a scan-line as sampled by the frame-grabber, and $s_{\text{rate}}$ is a sampling rate mismatch scale factor. The origin of the pixel coordinate frame is at the top-left corner of the digital image, with the $U, V$ axes along the columns and rows respectively.

The values of $K_{\text{ccd}}$, $\sigma_u$, $\sigma_v$ are obtained from the camera manufacturer's specifications, while the value of $K_U$ is obtained from the frame-grabber memory per scan-line. A previously developed camera calibration technique [149] is used to determine the remaining parameters: $C_u$, $C_v$, $f$, $\kappa$, $s_{\text{rate}}$, $R_c$, $\vec{t_c}$.

### 11.2.1.2    *Data Features*

The digital image, obtained from a *fixed* calibrated camera, is processed to extract vertex features, expressed in pixel coordinates. Given the intrinsic camera calibration parameters, it is straightforward to transform these pixel coordinates into image plane coordinates, expressed in the camera frame (Figure 11.3), by applying the pin-hole camera transformations in the reverse. We may regard this sequence of transformations of the raw image data, as a *data conditioning* step (Figure 7.2 on page 93), which results in image vertex features expressed in the *camera frame*:

$$\mathbf{z}^{(v)} \equiv \vec{r} = (u_c, v_c, f).$$

The set of such image vertex features $\{\vec{r}\}$ comprise the vision data features.

### 11.2.1.3  *Model Features*

The *visible* shape model vertices, expressed in the camera frame comprise the set of vision model features (Equation 11.4):

$$\mathbf{y}^{(v)} \equiv \vec{p}'_c = R_c \cdot \vec{p}' + \vec{t}_c, \qquad \vec{p}'_c \text{ is visible.}$$

The visibility is determined from the object geometry and the location of the viewpoint relative to it, both of which are known. Note that the *number* of model features therefore depends not only on the shape model, but also on the extrinsic camera calibration parameters, which determine the visibility.

### 11.2.1.4  *Constraint Equation*

Using a pin-hole camera model, an image vertex $\vec{r}$ is related to the corresponding shape model vertex $\vec{p}'_c$ by the perspective projection constraint of Equation (11.5). This perspective projection constraint may be written as:

$$\zeta \cdot \vec{r} = \vec{p}'_c = R_c \cdot \vec{p}' + \vec{t}_c$$

for some $\zeta \geq 1, \zeta \in \Re$, since $\vec{p}'_c$ lies on the ray joining the image vertex $\vec{r}$ to the lens center (Figure 11.3). This may be rewritten as,

$$R_c^T\{\zeta \cdot \vec{r} - \vec{t}_c\} = \vec{p}' = q\vec{p}q^* + \vec{t}$$

since $R_c^T R_c = I$. Simplifying, and noting that $\vec{e} = -R_c^T \cdot \vec{t}_c$ is the position of the lens center in the world coordinate frame [41], the vision perspective projection constraint equation is given by:

$$\vec{e} + \zeta \cdot R_c^T \cdot \vec{r} = q\vec{p}q^* + \vec{t} \qquad (11.6)$$

for some $\zeta \geq 1, \zeta \in \Re$.

We assume that at most one vision image vertex can correspond to a shape model vertex, i.e., the vision correspondence must be *one-to-one* (Section 6.2.2).

### 11.2.1.5   *Uncertainty Model*

The measurement space of the image data features is the finite image plane, which is subdivided into $K_U \times K_V$ rectangular pixels, each of size $\sigma_u \times \sigma_v$.

The sensor accuracy is given by approximating a rectangular pixel by an elliptical uncertainty region (Section 6.3.1.1, Figure 6.9), with the semi-principal axis lengths proportional to $\sigma_u$ and $\sigma_v$:

$$A = \left[ \begin{array}{cc} 1/\sigma_u^2 & 0 \\ 0 & 1/\sigma_v^2 \end{array} \right]$$

and radius $r = 1$.

The sensor precision is given the number of bits needed to address the measurement space (Section 6.3.1):

$$\lambda = \log_2(K_U \cdot K_V + 1).$$

### 11.2.1.6   *Representation Size*

Given shape model vertex feature, exactly one "ideal" data feature can arise from it, as given by the perspective projection constraint of Equation (11.5). This comprises the DCM point (Section 7.3.2.1), $\tilde{\vec{r}} = (\tilde{u}_c, \tilde{v}_c, f)$, given by:

$$\tilde{u}_c = \frac{f}{z_c} \cdot x_c, \qquad \tilde{v}_c = \frac{f}{z_c} \cdot y_c.$$

Since the DCM is a singular point, $\mathcal{L}[\tilde{\vec{r}} \,|\, \vec{p}\,'] = 0$.

The error residual representation size (Section 7.3.2.1) is calculated in the two-dimensional image plane, using an encoded accuracy and precision uncertainty model (Section 6.3.1):

$$\mathcal{L}[\vec{r} - \tilde{\vec{r}} \,|\, \tilde{\vec{r}}] \;=\; \mathcal{I}([u_c - \tilde{u}_c \quad v_c - \tilde{v}_c] \frac{A}{r^2} [u_c - \tilde{u}_c \quad v_c - \tilde{v}_c]^{\mathrm{T}})$$

where $n = 2$ in Equation (6.6).

The modeled data feature representation size is given by (Equation 7.15):

$$\mathcal{L}[\vec{r} \,|\, \mathbf{y} = \vec{p}\,'] = \mathcal{L}[\vec{\tilde{r}} \,|\, \vec{p}\,'] + \mathcal{L}[\vec{r} - \vec{\tilde{r}} \,|\, \vec{\tilde{r}}]$$
$$= \mathcal{I}([u_c - \tilde{u}_c \quad v_c - \tilde{v}_c] \frac{A}{r^2} [u_c - \tilde{u}_c \quad v_c - \tilde{v}_c]^{\mathrm{T}}). \tag{11.7}$$

The unmodeled data feature representation size (Section 7.3.2.1) is given by the vision sensor precision:

$$\mathcal{L}[\vec{r} \,|\, \mathbf{y}_0] = \lambda = \log_2(K_U \cdot K_V + 1). \tag{11.8}$$

### 11.2.2  Touch

#### 11.2.2.1  *Kinematic Calibration*

The hand-arm kinematic chain calibration is comprised of the **(a)** *actuator space* calibration which maps the joint sensor (potentiometers, encoders, etc.) readings into joint angles, and **(b)** Denavit-Hartenberg (D-H) parameters [41] which describe the link geometries and their relative positions in the kinematic chain. The measured joint angles are used with the D-H parameters to compute the location of a point on the finger-tip relative to the base of the robot.

#### 11.2.2.2  *Data Features*

A tactile sensor mounted on a finger-tip provides a contact location relative to the finger-tip. The forward kinematics of the hand-arm kinematic chain is used to compute the location of these contact points relative to the base of the robot. The set of such contact points $\{\vec{w}\}$ comprises the touch data features:

$$\mathbf{z}^{(t)} \equiv \vec{w}.$$

#### 11.2.2.3  *Model Features*

The shape model *faces*, $\{\mathbf{f}'\}$, comprise the touch model features:

$$\mathbf{y}^{(t)} \equiv \mathbf{f}'.$$

We assume that contact is made only on the object faces.

### 11.2.2.4 *Constraint Equation*

Since the fingers contact the object on its faces, an observed contact point data feature must be *contained* in a finite object face:

$$\vec{w} \cdot \vec{m}' = c', \qquad \vec{w} \in \mathbf{f}',$$

which simplifies to:

$$(\vec{w} - \vec{t}) \cdot (q\vec{m}q^*) = c, \qquad \vec{w} \in \mathbf{f}', \tag{11.9}$$

where the requirement $\vec{w} \in \mathbf{f}'$ asserts that $\vec{w}$ must lie within the finite polygonal face boundary. This constraint may be rewritten in parametric coordinates:

$$\vec{w} = q\{\vec{p}^a + \zeta(\vec{p}^b - \vec{p}^a) + \xi(\vec{p}^c - \vec{p}^a)\}q^* + \vec{t}, \qquad \vec{w} \in \mathbf{f}' \tag{11.10}$$

where $\zeta, \xi \in \Re$ define the location of $\vec{w}$, in the plane defined by the three non-collinear shape vertices $\vec{p}^a, \vec{p}^b, \vec{p}^c$.

Several contacts may be made on an object face, resulting in a *many-to-one* correspondence.

### 11.2.2.5 *Uncertainty Model*

The observation errors are specified in a *local* coordinate frame $\{X_t, Y_t, Z_t\}$ anchored at the contact location on the object surface, with the $Z_t$-axis coincident with the outward surface normal. An ellipsoidal uncertainty region (Section 6.3.1.1) is defined in this local contact coordinate frame, with semi-principal axis lengths given by $\sigma_{x_t}, \sigma_{y_t}, \sigma_{z_t}$ along the $X_t, Y_t, Z_t$ axes respectively:

$$A = \begin{bmatrix} 1/\sigma_{x_t}^2 & 0 & 0 \\ 0 & 1/\sigma_{y_t}^2 & 0 \\ 0 & 0 & 1/\sigma_{z_t}^2 \end{bmatrix}$$

and radius $r = 1$.

The measurement space of contact points is the finite three-dimensional region of points reachable by the hand-arm kinematic chain. Let us assume that it is a larger ellipsoid with the same shape as the uncertainty region ellipsoid, but with a radius $R$. The touch sensor precision is given by

$$\lambda = \log_2\left(\left[\frac{R}{r}\right]^3 + 1\right),$$

where $n = 3$ in Equation (6.7).

### 11.2.2.6 *Representation Size*

The DCM is the set of all points lying on the finite polygonal face:

$$\text{DCM}(\mathbf{f}') = \{\tilde{\vec{w}} : \tilde{\vec{w}} \in \mathbf{f}'\}.$$

The DCM is two-dimensional, $\tilde{n} = 2$, and its representation size is given by, $\mathcal{L}[\tilde{\vec{w}} \mid \mathbf{f}'] = \tilde{c}_\zeta + \tilde{c}_\xi = \tilde{c}_t$ (Equation 7.16), where $\tilde{c}_\zeta$ and $\tilde{c}_\xi$ is the fixed number of bits used to encode the parametric coordinates $\zeta$ and $\xi$ in the constraint Equation (11.10).

Given an observed touch data feature, $\vec{w}$, the "ideal" touch data feature, $\tilde{\vec{w}}$, is the point *closest* to the observed data feature, contained in the face $\mathbf{f}'$:

$$\tilde{\vec{w}} = \vec{w} - (\vec{w} \cdot \vec{m}' - c')\vec{m}'$$

where $(\vec{w} \cdot \vec{m}' - c')$ is the perpendicular distance between $\vec{w}$ and $\mathbf{f}'$. We check $\tilde{\vec{w}}$ for containment in $\mathbf{f}'$. For a $K$ sided *convex* polygonal face, with a boundary vertex list $< \vec{p}'_0, \ldots, \vec{p}'_{K-1} >$, ordered such that their traversal defines the face normal $\vec{m}'$ according to the right hand rule; a point $\tilde{\vec{w}}$ is contained iff:

$$[(\vec{p}'_k - \tilde{\vec{w}}) \times (\vec{p}'_{k+1 \mod K} - \tilde{\vec{w}})] \cdot \vec{m}' >= 0 \qquad \text{for all } k = 0, \ldots, K - 1.$$

Otherwise, if the above inequality is violated for the edge $< \vec{p}'_k, \vec{p}'_{k+1 \mod K} >$, we recompute the DCM point to be the one closest on this edge:

$$\tilde{\vec{w}} = \vec{p}'_k + \zeta(\vec{p}'_{k+1 \mod K} - \vec{p}'_k)$$

where

$$\zeta = \max(0, \min(1, \frac{(\vec{w} - \vec{p}'_k) \cdot (\vec{p}'_{k+1 \mod K} - \vec{p}'_k)}{(\vec{p}'_{k+1 \mod K} - \vec{p}'_k) \cdot (\vec{p}'_{k+1 \mod K} - \vec{p}'_k)}))$$

is a parametric coordinate along the edge $< \vec{p}'_k, \vec{p}'_{k+1 \mod K} >$.

The error residual representation size is computed in a local coordinate frame (Section 11.2.2.5) anchored at $\tilde{\vec{w}}$. The error $\Delta \vec{w} = \vec{w} - \tilde{\vec{w}}$ in world coordinates, is converted to the local coordinate frame by applying a rotation, $R_t$, which aligns the face normal $\vec{m}'$ to the $Z_t$-axis (i.e., to the vector

$(0, 0, 1))$:

$$\Delta \vec{w}_t = R_t \cdot \Delta \vec{w} = R_t \cdot (\vec{w} - \tilde{\vec{w}})$$

where,

$$R_t = \mathrm{ROT}(\vec{m}' \times (0, 0, 1), \arccos(\vec{m}' \cdot (0, 0, 1)))$$

is a rotation of angle $\arccos(\vec{m}' \cdot (0, 0, 1))$ about the axis $\vec{m}' \times (0, 0, 1)$. Therefore the error representation size is given by

$$\mathcal{L}[\vec{w} - \tilde{\vec{w}} \,|\, \tilde{\vec{w}}] = \mathcal{I}\big( \big[ (\vec{w} - \tilde{\vec{w}})^{\mathrm{T}} R_t{}^{\mathrm{T}} \frac{A}{r^2} R_t (\vec{w} - \tilde{\vec{w}}) \big]^{3/2} \big).$$

where $n = 3$, in Equation (6.6).

The modeled data feature representation size is given by (Equation 7.15):

$$\begin{aligned}
\mathcal{L}[\vec{w} \,|\, \mathbf{y} = \mathbf{f}'] &= \mathcal{L}[\tilde{\vec{w}} \,|\, \mathbf{f}'] + \mathcal{L}[\vec{w} - \tilde{\vec{w}} \,|\, \tilde{\vec{w}}] \\
&= \tilde{c}_t + \mathcal{I}\big( \big[ (\vec{w} - \tilde{\vec{w}})^{\mathrm{T}} R_t{}^{\mathrm{T}} \frac{A}{r^2} R_t (\vec{w} - \tilde{\vec{w}}) \big]^{3/2} \big). \quad (11.11)
\end{aligned}$$

The unmodeled data feature representation size (Section 7.3.2.1) is given by the touch sensor precision:

$$\mathcal{L}[\vec{w} \,|\, \mathbf{y}_0] = \lambda = \log_2\big( \big[ \frac{R}{r} \big]^3 + 1 \big). \quad (11.12)$$

### 11.2.3  *Grasp*

#### 11.2.3.1  *Kinematic Calibration*

The kinematic calibration required for the grasp data is the same as that for the touch sensor, and is described in Section 11.2.2.1 on page 181.

#### 11.2.3.2  *Data Features*

A *raw* grasp data feature obtained from tactile sensors mounted on the finger-tips consists of a contact point location, $\vec{w}$, and a local surface normal, $\vec{n}$, pointing outwards. The hand-arm forward kinematics is used to compute these quantities in the world frame.

The raw grasp data is *conditioned* to compute planar grasp data faces, {s}, which comprise the grasp data features. A grasp data face feature s is given by:

$$z^{(g)} \equiv s = (\vec{n}, d)$$

expressed in a coordinate frame obtained by shifting the world frame to the centroid of the raw contact points:

$$< \vec{w} > = \frac{1}{N} \sum_{i=1}^{N} \vec{w}_i.$$

The perpendicular distance of the data face from the origin, $d$, is given by:

$$d = \vec{n} \cdot (\vec{w} - <\vec{w}>)$$

where $\vec{n}$ is the observed *outward* surface normal.

The world coordinate frame is shifted to the centroid of the raw contact points, $<\vec{w}>$, to reduce the effect of compounded errors in computing $d$ (since it is a *product* of two measured quantities).

### 11.2.3.3  *Model Features*

The shape model *faces*, {f'}, comprise the grasp model features:

$$y^{(g)} \equiv f'.$$

### 11.2.3.4  *Constraint Equation*

Ideally, and observed data face, s, must coincide exactly with a model face, f':

$$\vec{n} = \vec{m}'$$
$$d = c' - \vec{m}' \cdot <\vec{w}> \qquad (11.13)$$

where the model face is shifted to the raw data centroid frame.

We assume that at most one data face may arise from a model face, i.e., the grasp correspondence must be *one-to-one*.

### 11.2.3.5  *Uncertainty Model*

The observation errors in measuring a data face, are specified by the errors in the surface normal and the errors in the perpendicular distance from the origin. The observation error in the surface normals, given by the angle between them, is specified by a one-dimensional angular uncertainty region (Section 6.3.1), with an angular quantization interval $\delta_{\angle}$. The observation error in the distance from the origin is specified by another one-dimensional uncertainty region, with a quantization interval $\delta_{\perp}$.

The measurement space of the grasp sensor is the finite region accessible by the hand-arm kinematic chain. Let the farthest planar face in this measurement space be at a distance $R$ from the origin. The number of bits needed to address this space of "distances from the origin" is given by $\log_2(R/\delta_{\perp} + 1) + 1$, where the extra one bit is added for the sign. The space of surface normals can be addressed by specifying a point on the unit sphere. Using polar coordinates, exactly two angles (namely $\varphi$ and $\psi$ in Figure 11.1) must be specified, requiring $\log_2(2\pi/\delta_{\angle} + 1) + \log_2(\pi/\delta_{\angle} + 1)$ bits. Therefore, the grasp sensor precision is given by:

$$\lambda = \log_2\left(\frac{2\pi}{\delta_{\angle}} + 1\right) + \log_2\left(\frac{\pi}{\delta_{\angle}} + 1\right) + \log_2\left(\frac{R}{\delta_{\perp}} + 1\right) + 1.$$

### 11.2.3.6  *Representation Size*

The ideal data face, $\tilde{s}$, which can arise from a model face, is given by the model face itself, shifted to the raw data centroid frame:

$$\tilde{n} = \vec{m}'$$
$$\tilde{d} = c' - \vec{m}' \cdot <\vec{w}>$$

and comprises the grasp DCM. Since the DCM contains a single element, it is 0 dimensional, and its representation size, $\mathcal{L}[\tilde{s} \,|\, \mathbf{f}'] = 0$.

The error between an observed grasp data face feature and an ideal DCM face feature, $\Delta s = (\vec{n} - \tilde{n}, d - \tilde{d})$, is the "distance" between two planar faces in the space of all planes. It consists of the error between the surface normals, and the error in the distances from the origin. These two components are encoded independently, such that given $\tilde{s}$, the observed data feature, $s = (\vec{n}, d)$, may be regenerated.

Given $\tilde{n}$, the data face normal, $\vec{n}$, can be regenerated as a rotation about the axis $\tilde{n} \times \vec{n}$ by an angle $\arccos(\tilde{n} \cdot \vec{n}) \in [0, \pi]$. The unit vector $\tilde{n}$ may be regarded at the *pole* of a new spherical (polar) coordinate system. In this new spherical coordinate system, the axis of rotation, $\tilde{n} \times \vec{n}$, is specified by an angular position on the equator ($\varphi$ in Figure 11.1), and can be encoded in $\log_2(2\pi/\delta_\angle + 1)$ bits. The angle of rotation is the deviation from the pole ($\psi$ in Figure 11.1), and can be encoded in $\mathcal{I}(\lfloor \arccos(\tilde{n} \cdot n)/\delta_\angle \rfloor)$ bits.

The error in the distances from the origin can be encoded in

$$\mathcal{I}(\lfloor |\tilde{d} - d|/\delta_\perp \rfloor) + 1$$

bits, where the extra one bit is added for the sign.

The modeled data feature representation size is given by (Equation 7.15):

$$\begin{aligned}
\mathcal{L}[\mathbf{s} \,|\, \mathbf{y} = \mathbf{f}'] &= \mathcal{L}[\tilde{\mathbf{s}} \,|\, \mathbf{f}'] + \mathcal{L}[\mathbf{s} - \tilde{\mathbf{s}} \,|\, \tilde{\mathbf{s}}] \\
&= \log_2(\frac{2\pi}{\delta_\angle} + 1) + \mathcal{I}(\frac{\arccos(\tilde{n} \cdot n)}{\delta_\angle}) + \mathcal{I}(\frac{|\tilde{d} - d|}{\delta_\perp}) + 1.
\end{aligned} \tag{11.14}$$

The unmodeled data feature representation size (Section 7.3.2.1) is given by the grasp sensor precision:

$$\mathcal{L}[\mathbf{s} \,|\, \mathbf{y}_0] = \lambda = \log_2(\frac{2\pi}{\delta_\angle} + 1) + \log_2(\frac{\pi}{\delta_\angle} + 1) + \log_2(\frac{R}{\delta_\perp} + 1) + 1. \tag{11.15}$$

## 11.3   Shape Matching from Grasp

In this section we consider the problem of shape matching and pose estimation, using only the grasp sensor. A robot hand mounted on a robot arm explores the environment by grasping. Tactile sensors mounted on the hand record the coordinates of the contact point and the surface normals, as shown in Figure 11.4. This application emphasizes interpretation estimation for a *fixed* (immobile) rigid object, using multiple grasps to *explore* the object. We assume that sufficient number grasp data features are acquired so that the data set is complete. The grasp sensor model, developed in Sections 11.2.3 is used with the object models described in Section 11.1.

### 11.3.1   *Constraining Data Feature Sets*

A grasp (G) constraint (Equation 11.13) partially constrains the object orientation and position, so that a model face coincides with an observed data face. Three such constraints may used to completely define the object pose. The resulting CDFS is a called a GGG, also referred to as a *3-on-3* transform. The 3-on-3 transform *aligns* three arbitrary model face features $\mathbf{f}'_{j_1}$, $\mathbf{f}'_{j_2}$, $\mathbf{f}'_{j_3}$ with three arbitrary data face features $\mathbf{s}_{i_1}$, $\mathbf{s}_{i_2}$, $\mathbf{s}_{i_3}$

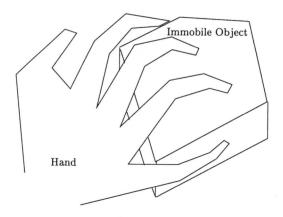

Fig. 11.4   The hand explores a fixed rigid object by grasping multiple times, until a sufficient number of grasp data features are obtained. The grasp data features are used to estimate the position and orientation of the object.

respectively, as shown in Figure 11.5, This system of constraint equations is *over-constrained*, and can only be satisfied partially, to compute a pose transform as discussed in Appendix G.

Each CDFS system of equations may result in at most one solution. Therefore the total number of CDFS poses is given by

$$\binom{N^{(g)}}{3}\binom{M^{(g)}}{3} 3! \tag{11.16}$$

where $N^{(g)}$ and $M^{(g)}$ are the number of grasp data and model features respectively.

### 11.3.2   *Shape Matching Algorithm*

The polynomial time hypothesize and test algorithm, described in Section 8.4, can be used to find the minimal representation size interpretation from grasp data. A *two-part* encoding scheme is used and the one-to-one correspondence is computed as described in Section 8.3.1. The total time complexity is given by:

$$O(M^3 \cdot N^3) \times [O(M \cdot N) + O(M \cdot N \cdot \min(M, N))]$$
$$= O(M^4 \cdot N^4 \cdot \min(M, N)).$$

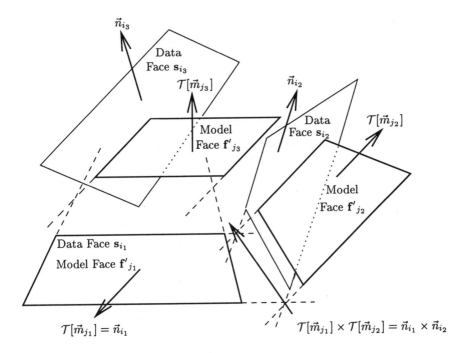

Fig. 11.5 A 3-on-3 transform aligns 3 arbitrary model features $\mathbf{f'}_{j_1}$, $\mathbf{f'}_{j_2}$, $\mathbf{f'}_{j_3}$ with 3 arbitrary data features $\mathbf{s}_{i_1}$, $\mathbf{s}_{i_2}$, $\mathbf{s}_{i_3}$ respectively.

We could also use a differential evolution (DE) program, described in Section 8.5.3.2 on page 142. The pose parameter representation is as described in Section 11.1.1. For typical problems with $N, M \approx 10$, the hypothesize and test algorithm is preferred, since it guarantees finding the minimal representation size interpretation in a reasonable time.

### 11.3.3 *Computer Simulations*

Simulated data was generated as follows. The object model is described in model coordinates, and generates a set of "perfect" data features from it. The simulated data is generated from perfect data features by removing a random subset of features, and adding a set of randomly generated features. The perfect data features that remained are perturbed by adding noise to them. A randomly-generated zero-mean Gaussian noise vector is added to the contact point, resulting in an uncertainty *sphere* around the contact point. The face surface normal unit vector is perturbed by a randomly generated zero mean Gaussian angle from its nominal direction, resulting in an uncertainty *cone* about the face normal.

The shape matching program has no knowledge of the data generation, and is given only the model description and the simulated data. Its output is the optimal pose and correspondence. In this manner the performance of the algorithm can be studied under varying amounts of noise and missing/spurious features in the data. The outcomes of two such simulated experiments are tabulated in Table 11.2 for one of the models used to test the algorithm. The correct correspondences were found, even when more than half of the data features are spurious, and the data features arising from the model are noisy.

### 11.3.4  *Laboratory Experiments*

A finger of the *Anthrobot* hand [4] (Figure 12.3) was used to gather the grasp data for the pose estimation experiments. The kinematic model of the hand was calibrated experimentally using the *Bird* [219] position and orientation sensor mounted on the wrist of the hand. A force sensing resistor mounted on the finger-tip, with a compliant pad over it, serves to detect when a contact has been made. The finger-tip mounted sensor has a cone of hand orientations for which it registers a valid resistance reading, when in contact with a planar surface. Therefore the mounted sensor reading depends on the amount of force and its distribution. The joints of the finger are kept in a fixed posture under PID control throughout the experiment.

The kinematics of the fixed posture hand is calibrated between the point where the finger-tip sensor registers a minimum resistance, and the point where the *Bird* sensor is mounted on the wrist. The hand is moved on the rigid object surface until the tactile sensor registers a minimum resistance reading and the corresponding finger-tip sensor position and orientation are computed. This experimental approach allows us to collect the tactile data needed for the shape matching experiment.

The results of one such matching experiment are displayed in Figure 11.6. The reference pose, shown in Figure 11.6, was obtained experimentally using the *Bird* sensor directly to measure the location and surface normal at the contact points. The estimated pose, calculated using the matching algorithm, is shown superimposed on the reference pose in Figure 11.6.

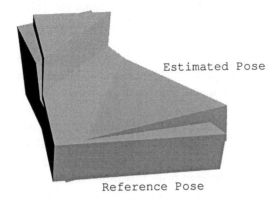

Estimated Pose

Reference Pose

Fig. 11.6    Matching results on grasp data obtained by probing the *Pedestal* object (see Figure 11.8 on page 194) with an *Anthrobot* finger. In this experiment 8 contact points were obtained, 7 of which were collected by contacting 7 distinct faces of the 8-faced *Pedestal* object; the remaining contact point was spurious. The estimated pose of the object is shown superimposed on the reference pose. The position deviation for this experiment was less than 5%, and the orientation error was $7.2°$ in roll, $-3.2°$ in pitch, and $-7.1°$ in yaw.

## 11.4    Object Recognition by Fusing Vision and Touch

In this section we consider the original tactile–visual fusion problem of Figure 1.1 on page 2, where the vision and touch sensors must be fused to estimate the identity and pose of an object being held by the robot hand. This application emphasizes interpretation estimation *during* grasp, where multiple contact points are obtained simultaneously, while the camera takes an image of the hand holding the object. The touch data may be noisy and incomplete due to required hand configurations for a stable grasp, while the vision data may contain spurious features arising from the presence of the hand in the image. The vision and touch sensor models, developed in Sections 11.2.1 and 11.2.2 are used with the object models described in Section 11.1.

### 11.4.1    *Constraining Data Feature Sets*

The position and orientation (a total of 6 parameters) of a given polyhedral shape are hypothesized using image vertex features from the vision sensor and contact point features the touch sensor.

The vision (V) constraint equation of Equation (11.6) gives 3 (real) equations in 7 unknowns (6 pose parameters, and $\zeta$); the touch (T) constraint equation of Equation (11.9) gives 1 equation in 6 unknowns (the 6 pose parameters). Accordingly the CDFS combinations are: VVV using three vision constraints, VVTT using two vision and two touch constraints, VTTTT using one vision and four touch constraints, and TTTTTT using six touch constraints.

In general, each CDFS system of equations may be solved partially or exactly, and results in multiple pose solutions as described in Appendix I. The maximum number of solution for each type of CDFS are summarized in Table 11.3. As can be seen from this table, multiple solutions exist for each CDFS. The total number of CDFS poses is given by:

$$8 \binom{N^{(v)}}{3} \binom{M^{(v)}}{3} 3! + 64 \binom{N^{(v)}}{2} \binom{M^{(v)}}{2} 2! \binom{N^{(t)}}{2} \binom{M^{(t)}}{2} 2! +$$

$$512 \binom{N^{(v)}}{1} \binom{M^{(v)}}{1} \binom{N^{(t)}}{4} \binom{M^{(t)}}{4} 4! + 4096 \binom{N^{(t)}}{6} \binom{M^{(t)}}{6} 6! \quad (11.17)$$

where $N^{(v)}$ and $M^{(v)}$ are the number of vision data and model features respectively, while $N^{(t)}$ and $M^{(t)}$ are the number of touch data and model features respectively.

### 11.4.2    *Multisensor Object Recognition Algorithm*

The polynomial time hypothesize and test algorithm, described in Section 8.4 on page 132, requires evaluating all possible CDFS. Evaluating a CDFS requires solving a system of polynomial equations, as described in Appendix I, and the total number of such evaluations is $O(N^6 M^6)$, a high order polynomial. In practice, solving one of these CDFS system of polynomial equations can take a few seconds (on a SPARC 20), and therefore this method is not computationally practical (also see Section 10.5.2), for most problems of interest ($N, M \approx 10 - 20$).

Instead, we use a differential evolution (DE) program, described in Section 8.5.3.2. The pose parameter representation is described in Section 11.1.1. A two-part encoding scheme is used for both sensors and the data correspondences are computed as described in Section 8.3.1. In practice, the DE finds a minimal representation size interpretation in a few minutes (on a SPARC 20) for most problems of interest (Section 11.5.2).

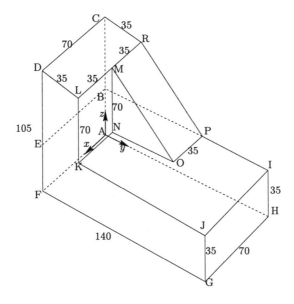

Fig. 11.7   The *Lwedge* object in its model coordinate frame.

## 11.5   Computer Simulations

In this section we present computer simulation results for the three-dimensional multisensor fusion problem. Detailed laboratory experimental results are presented in Chapter 12. The three-dimensional object shapes used in computer simulations and laboratory experiments are shown in Figures 11.7 and 11.8. All the dimensions are in *millimeters*.

To generate a simulated environment model, an object shape is chosen randomly from the environment model library, and its parameters are instantiated randomly, $\varphi \in [-\pi, \pi]$, $\cos \psi \in [0, 1]$, $\phi \in [0, 2\pi]$, $t_x \in [980, 1030]$, $t_y \in [80, 130]$, $t_z \in [-145, -95]$. A camera with a focal length of 24.56 is used to generate the simulated vision data. The camera extrinsic parameters are given by a rotation matrix:

$$R_c = \begin{bmatrix} +2.365478 \times 10^{-2} & -9.990603 \times 10^{-1} & -3.631681 \times 10^{-2} \\ -6.367009 \times 10^{-3} & -3.647679 \times 10^{-2} & +9.993142 \times 10^{-1} \\ -9.996999 \times 10^{-1} & -2.340733 \times 10^{-2} & -7.223876 \times 10^{-3} \end{bmatrix}$$

and position $\vec{t}_c = (120.7553, 143.0887, 2000.1075)$. These settings are the

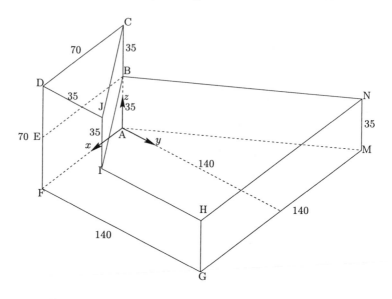

Fig. 11.8   The *Pedestal* object in its model coordinate frame.

same as those for the laboratory Experiment 1 (see Section 12.5), and are typical of real-life scenarios.

The vision sensor was modeled by a two-dimensional ellipsoidal uncertainty region of with the semi-principal axis lengths $8.3333 \times 10^{-3}$ and $9.7166 \times 10^{-3}$ respectively (Figure 6.9 on page 85), and a measurement space of radius 3.2. These values are typical of a $640 \times 480$ image obtained from a 1/2" CCD camera. The touch sensor was modeled by a three-dimensional ellipsoidal uncertainty region of with the semi-principal axis lengths equal to 0.1 along each axis, and a measurement space of radius 150. These values are typical of a hand-arm manipulator positional accuracy and workspace volume. These vision and touch uncertainty models were used to randomly generate the multisensor data from the randomly generated environment model.

A one-to-one correspondence model was used for the vision sensor, while a many-to-one correspondence model was used for the tactile sensor. In both cases, a two-part fixed-length encoding algorithm (Section 7.3.2.2) was used to compute the representation size, using the sensor models developed in Sections 11.2.1 and 11.2.2.

## 11.5.1 *Model Selection Performance*

The model selection performance of the minimal representation size multisensor technique was evaluated on 100 randomly generated problems. A typical problem size was chosen, consisting of 16 vision data features of which 50% were spurious, and 9 tactile data features of which 15% were spurious.

A 100 member DE/best/2 differential evolution algorithm with F = 0.8, and CR = 0.8 (Sections 8.5.2 and 8.5.3.2), was used to search for the minimal representation size pose and the correspondences for each sensor. The search is terminated when the population reaches 50% convergence (parameter tolerance of 0.1), or when the maximum time limit of 900 CPU seconds is exceeded, or when a maximum of 1000 generations have been completed. The search is considered to have correctly converged (to the global minimum), if it reaches within 14 bits of the simulated multisensor data representation size.

The model selection performance of the technique is assessed from the number of misclassifications, and the average pose and correspondence errors among those correctly classified. The pose error is given by the angle between the rotational axes: $\hat{n}\angle n$, the error in the rotation angle: $\hat{\phi} - \phi$ mod $2\pi$, and the translation error: $|\hat{\vec{t}} - \vec{t}|$, where $\hat{x}$ denotes the estimated value of $x$. The correspondence error is given by the number of misassignments in the estimated correspondence.

A typical simulation sample is shown in Figure 11.9. The simulation results on 100 randomly generated problems for an environment library containing two object shapes, *Lwedge* and *Pedestal*, are tabulated in Table 11.4 on page 201. The mean errors and standard deviations are tabulated among those DE searches which correctly converged, and the correct object shape was chosen.

## 11.5.2 *Search Performance*

The time complexity of the differential evolution (DE) search algorithm was evaluated on different problem sizes. The environment model library had exactly one environment model: the *Lwedge* object. A 100 member DE/best/2 differential evolution algorithm with F = 0.8, and CR = 0.8 was used to search for the minimal representation size pose and the correspondences for each sensor. The search algorithm was run on randomly

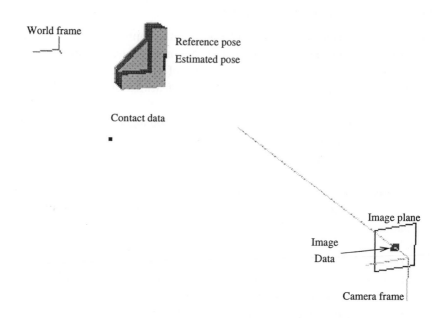

Fig. 11.9   Simulation results for the *Lwedge* object, showing the reference pose superimposed on the estimated pose found by fusing vision and touch data.

generated problems with varying number of vision and touch data features. The DE search is terminated when the population reaches 50% convergence (parameter tolerance of 0.1), or when the maximum time limit of 900 CPU seconds is exceeded, or when a maximum of 1000 generations have been completed. Results are plotted for the DE runs which correctly converged (reached within 14 bits of the simulated multisensor data representation size).

Figure 11.10 shows the CPU time against the number of vision data features, for a fixed number of touch data features. Figure 11.11 shows the number of minimal representation size correspondence evaluations (Section 8.3) against the number of vision data features. Figure 11.12 shows the CPU time against the number of touch data features, for a fixed number of vision data features. Figure 11.13 shows the number of minimal representation size correspondence evaluations against the number of touch data features.

From these plots, we see that the DE converges in approximately 200

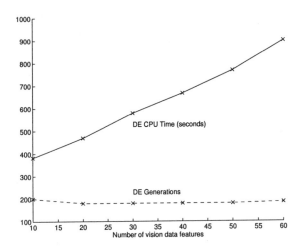

Fig. 11.10 CPU time (SPARC 20) vs. the number of vision data features for the differential evolution (DE) search algorithm. The vision sensor had 50% outliers, while the touch sensor produced 10 data features with 15% outliers.

generations, and requires approximately 16000 minimal representation size correspondence evaluations. These values seem to change very little with the problem size. The CPU time increases linearly with the number of data features $N$. This also follows from the theoretical analysis (see Section 8.3.1), since the time taken to compute a one-to-one vision minimal representation size correspondence grows as $O(M \cdot N \min(M, N))$ which is linear in $N$ when $M \leq N$, while the time taken to compute a many-to-one touch correspondence grows as $O(M \cdot N)$ correspondence!many-to-one which is linear in $N$. In fact, as the number of data features increases, the DE appears to converge correctly more consistently. This may be due to the fact that the global minimum is deeper as the number of data features increases. As seen from Table 11.4, 81% of the DE searches correctly converged to the global minimum. This number may be further improved by tuning the DE search parameters $P$, $F$, and CR. In practice, several DE searches may be executed in (in parallel), to improve the reliability of the search.

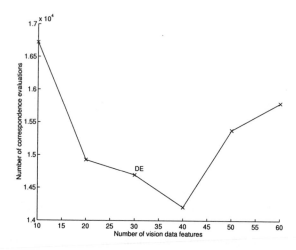

Fig. 11.11  Number of minimal representation size correspondence evaluations vs. the number of vision data features for the differential evolution (DE) search algorithm.

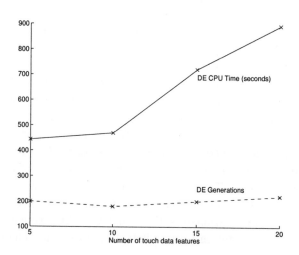

Fig. 11.12  CPU time (SPARC 20) vs. the number of touch data features for the differential evolution (DE) search algorithm. The touch sensor had 15% outliers, while the vision sensor produced 20 data features with 50% outliers.

Table 11.1  Environment and sensor models in three-dimensional object recognition.

| Symbol | Meaning |
|---|---|
| | *Environment Model* |
| $\Xi$ | polyhedral shape structure: geometry described in local object coordinates |
| $\theta$ | pose $\mathcal{T}[\cdot]$: rotation $R$, and translation $\vec{t}$ |
| $q_E$ | polyhedral shape model: $q_E = \Xi(R, \vec{t}) \equiv \mathcal{T}[\Xi]$ |
| $Q_E$ | shape library: set of objects under consideration in all possible poses |
| | *Vision Sensor (V)* |
| $\Psi$ | camera accuracy is the pixel size |
| $\Upsilon$ | measurement space is the camera image size |
| $\alpha$ | camera calibration: intrinsic (focal length $f$, image center, lens distortion, sampling rate mismatch) and extrinsic (world position, $\vec{t}_c$, and orientation, $R_c$ in camera frame) |
| $z$ | image vertex features in camera frame, $\vec{r} = (u_c, v_c, f)$ |
| $y$ | *visible* shape vertices in camera frame, $\vec{p}\,'_c = R_c \cdot \mathcal{T}[\vec{p}\,] + \vec{t}_c$ |
| $\omega$ | one-to-one correspondence between image point features and model vertices |
| $h$ | perspective vision constraint: $\zeta \vec{r} = \vec{p}\,'_c$ |
| $\mathcal{F}$ | *visible* vertex feature computation from shape model, $\{\vec{p}\,'_c\} = \mathcal{F}(\Xi(R, \vec{t}); \vec{t}_c, R_c)$ |
| | *Touch Sensor (T)* |
| $\Psi$ | touch sensor accuracy |
| $\Upsilon$ | measurement space is the finite region accessible by the hand-arm kinematic chain |
| $\alpha$ | hand-arm kinematics |
| $z$ | finger-tip contact point locations, $\vec{w}$, computed using the robot arm-hand kinematics |
| $y$ | polygonal faces of the shape model, $\mathbf{f}' = (\vec{m}', c')$, specified by a face normal $\vec{m}'$, and distance from the origin $c'$ |
| $\omega$ | many-to-one correspondence between contact points and model faces |
| $h$ | a contact point must lie on a model face: $\vec{w} \cdot \vec{m}' = c'$ |
| $\mathcal{F}$ | model face extraction from shape model, $\{\mathbf{f}'\} = \mathcal{F}(\Xi(R, \vec{t}); )$ |
| | *Grasp Sensor (G)* |
| $\Psi$ | grasp sensor accuracy |
| $\Upsilon$ | measurement space is the finite region accessible by the hand-arm kinematic chain |
| $\alpha$ | hand-arm kinematics |
| $z$ | planar face, $\mathbf{s} = (\vec{n}, d)$, where $\vec{n}$ is the face normal, and $d$ is the distance from origin |
| $y$ | planar face, $\mathbf{f}' = (\vec{m}', c')$, specified by a face normal $\vec{m}'$, and distance from the origin $c'$ |
| $\omega$ | one-to-one correspondence between data faces and model faces |
| $h$ | an observed data face must coincide with a model face: $\mathbf{s} = \mathbf{f}'$ |
| $\mathcal{F}$ | model face extraction from shape model, $\{\mathbf{f}'\} = \mathcal{F}(\Xi(R, \vec{t}); )$ |
| $\zeta, \xi$ | intermediate variables |

Table 11.2   Matching performance on simulation data for a solid cuboid object with varying amounts of noise and missing and spurious features.

|  | Clean | Noisy |
|---|---|---|
| *Data Generation* | | |
| Number of model features, $M$ | 6 | 6 |
| Number of data features, $N$ | 4 | 9 |
| Number of modeled data features | 4 | 4 |
| Contact point noise std. dev. | 0.00 | 0.10 |
| Surface normal noise std. dev.(degrees) | 00.0 | 15.0 |
| *Matching Results* | | |
| Number of data features modeled | 4 | 5 |
| Number of correct correspondences | 4 | 4 |

Table 11.3   Computing three-dimensional pose from vision and touch data.

| CDFS | Unknowns | Extra Variables | Quadratics | Solutions |
|---|---|---|---|---|
| VVV | 9 | 0 | 3 | 8 |
| VVTT | 8 | 4 | 6 | 64 |
| VTTTT | 7 | 8 | 9 | 512 |
| TTTTTT | 6 | 12 | 12 | 4096 |

Table 11.4   Model selection performance an environment model library containing two object models, on 100 randomly generated problems. Each problem had 16 vision data features with 50% outliers, and 9 touch data features with 15% outliers.

|  | Environment Model Library<br>*Lwedge*<br>*Pedestal* |
| --- | --- |
| Trials | 100 |
| Correctly Converged | 81 |
| Misclassified | 0 |
| Rotation axis error, mean | 0.033372° |
| standard deviation | 0.184239° |
| Rotation angle error, mean | 0.028648° |
| standard deviation | 0.165689° |
| Translation error, mean | 0.165613 |
| standard deviation | 0.100419 |
| Vision correspondence error, mean | 2.222222 |
| standard deviation | 1.151086 |
| Touch correspondence error, mean | 0.172840 |
| standard deviation | 0.380464 |
| Representation size deviation, mean | -6.340123 |
| standard deviation | 3.828411 |

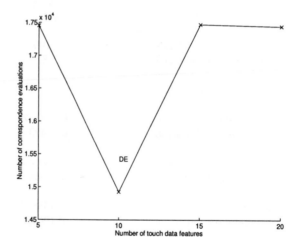

Fig. 11.13   Number of minimal representation size correspondence evaluations vs. the number of touch data features for the differential evolution (DE) search algorithm.

# Chapter 12

# Laboratory Experiments

In this chapter we describe laboratory experiments which demonstrate the minimal representation size technique on tactile–visual multisensor fusion. These experiments serve as a "proof-of-concept" of the technique developed in this book, and demonstrate that it is a practical method for solving multisensor fusion problems.

## 12.1  Introduction

Figure 1.1 on page 2 shows a robot hand holding an object in view of a camera. In these experiments, the finger-tips of the hand sense the approximate contact positions on the planar object surfaces, while the camera detects vertices of the same object. The objective of the experiments is to utilize a combination of tactile and vision data to determine the object pose.

As shown in Figure 1.1, the hand is visible in the field-of-view of the camera, and introduces extraneous vertex features in the image. Typically, about half of the image vertex features are due the the hand occluding parts of the object—thus the vision data alone may not be sufficient to estimate the three-dimensional pose of the object. We used a five-fingered hand, and for typical grasps, at most three distinct object surfaces are contacted by the finger-tips. The tactile data from these contacts is usually not rich enough to uniquely estimate the three-dimensional object pose by itself.

By fusing the vertex features from the image and the contact points from the finger-tips, we expect to correctly estimate the object pose and also the feature correspondence in the two data sets. We expect the fused estimates to be no worse than those obtained by using either sensor alone; and, in

Fig. 12.1   A picture of the laboratory experimental setup. The hand is mounted on a *PUMA* arm, and is used to hold an object in the field-of-view of a fixed camera.

practice, the results described below indicate a dramatic improvement in the fused estimates over those from individual sensors.

## 12.2   Experimental Setup

The experimental setup is shown in Figure 12.1, and a block diagram of the system is shown in Figure 12.2. A five-fingered *Anthrobot-3* [4] robot hand (Figure 12.3) was mounted on a six degree-of-freedom (DOF) articulated *PUMA-760* [199] robot arm. The motion of the hand was controlled via teleoperation (see Figure 12.2) using a *Cyberglove* [204]. The human operator wears the Cyberglove and the hand mimics the movements of the glove. The arm movements are controlled using the teach pendant supplied with the robot arm, via the *VAL II* [200] control language. For each experiment, the hand was placed in the field-of-view (FOV) of a calibrated camera (Section 12.3.2), with the palm facing upwards. A polyhedral object was placed on the palm, and the fingers flexed in teleoperation mode, to grasp the object.

In order to provide tactile sensing capability, special finger-tips with tactile sensors were fabricated for the *Anthrobot-3* hand. The tactile sen-

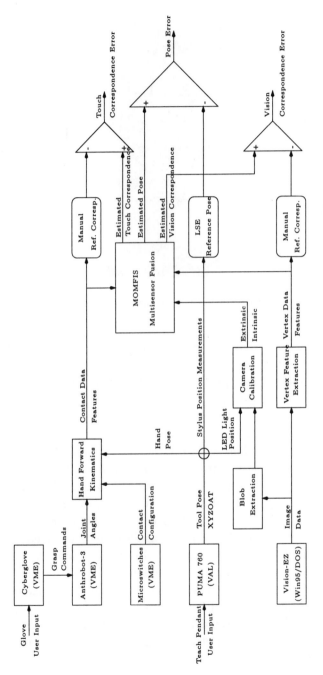

Fig. 12.2   Block diagram of the laboratory experiments.

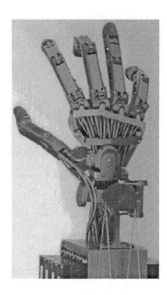

Fig. 12.3   *Anthrobot-3*, a five-fingered anthropomorphic robot hand.

sors on each finger-tip consisted of three *Microswitch* [137] micro-switches mounted together, to form a triangle. The on/off state of the three switches on a finger-tip specify the type of contact—for a parallel planar contact all the three switches are activated. The switch locations on the finger-tips were carefully measured, so that their contact positions could be computed with respect to the *PUMA* base frame.

The camera was a *Pullnix TM-7EX*, mounted on a rigid but adjustable fixture, and placed directly in front of the robot, about 2m away (see Figure 12.1). A *Cosmicar* 25mm focal length lens was used, with a small aperture setting ($< f/16$), and focussed at infinity. A small aperture approximates the *pin-hole* model used for camera calibration. The objects were painted white to create a contrast against the black hand. A black backdrop was used to prevent background objects in the lab from appearing in the camera view and to minimize reflections.

For the purpose of taking reference measurements for evaluation of the results, an adjustable, but rigid fixture was used to clamp the object held by the hand, such that it would be "frozen" in its pose. The reference measurements consisted of object vertex coordinates, obtained by probing the object using a *stylus* attached to the *PUMA* mounting flange.

The computing hardware (see Figure 12.2) consisted of a real-time four processor VME cage running the VxWorks operating system, a Pentium PC running Windows95, and Sun workstations running Unix (Solaris); all of these share a common local-area-network. The tactile sensors, the hand controller, and the glove-hand teleoperation ran on the VME cage, while the *Vision-EZ (DT55)* framegrabber was located on the PC. A distributed application collected multisensor data and stored it on the shared NFS file system. The feature extraction and reference pose computation were done on the Sun workstations.

The real-time software for controlling the hand and teleoperation, used the CATECHISM framework [96]. A locally ported version of the POP-EYE [23] image processing software was used for image processing and image vertex feature extraction. A locally developed software package [149], was used for camera calibration. Another package for the hand kinematics [164, 180] was adapted for this experimental setup. The *PUMA* control software ran in its own stand-alone controller, and was programmed in the *VAL II* robot programming language.

The **MO**del-based **M**ultisensor **F**usion and **I**nterpretation **S**ystem (MOMFIS) software package (see Appendix B), developed as part of this work, is an extensible object-oriented application framework implemented in the *C++* programming language. It was used to process the multisensor data, and evaluate the ideas developed in this work.

## 12.3 Calibration Procedures

The procedures used to calibrate the sensors are described next. In these experiments, we assume that the *PUMA base frame is the world frame*. All the *dimensions are in millimeters*, unless specified otherwise.

### 12.3.1 *Arm and Hand Calibration*

The kinematic parameters of the *PUMA* arm and the joint angle calibration parameters are supplied by the manufacturer. On startup, we loaded the calibration file, and ran the `CALIBRATE` command at the *VAL* prompt, to self calibrate the *PUMA* arm. Since the *PUMA* arm is used for positioning during camera calibration, tactile sensor measurements, and reference pose measurements, their accuracy is limited by the accuracy of this *PUMA* calibration. Since the object reference pose is also determined by the *PUMA* arm, the *repeatability* of the arm is the most relevant measure. The *PUMA* manual claims a repeatability of ±0.2mm.

The *Anthrobot-3* joint potentiometers were calibrated by measuring their reading for different joint angles, which were measured with a protractor. The kinematic parameters were available from prior work [164, 180], and are based on the design dimensions of the hand. These kinematic parameters are used to compute the finger-tips contact locations with respect to the *PUMA* base frame. The computed contact locations were further corrected to account for stretching of the tendons in the hand. These corrections were based on determining the locations of marked finger-tip positions in the reference pose of the object. The accuracy of the contact locations obtained in this manner is ±2mm.

### 12.3.2   *Camera Calibration*

As discussed in Section 11.2.1.1 on page 177, the camera's five *intrinsic* parameters comprising of the radial lens distortion coefficient $\kappa$, the focal length $f$, the sampling rate mismatch $s_{\text{rate}}$, and the image center $(C_u, C_v)$; and the six *extrinsic* parameters comprising of the position $\vec{t}_c$ and orientation $R_c$ of the world frame in the camera frame must be calibrated. A previously developed method [149] was used, which requires a series of world coordinates and the image pixel coordinates of a point in space.

The data for camera calibration was collected by sampling a cuboidal region of space with a point light source. A specially fabricated plate with a bright light LED in the center, was attached to the *PUMA* mounting flange. A *VAL* program was written for the PUMA to automate this procedure. For each position of the light source, its image was taken, and the centroid of the largest blob [43] was computed (the room lights had to be turned off to ensure that the light source was the largest blob in the image). This centroid was taken to be the image coordinates of the light source, while its world coordinates were obtained by querying the *PUMA*. A total of 64 points were collected by uniformly sampling a 150x100x200 cuboidal region. The final distance to ray root-mean-square (RMS) error [149] in the camera calibration was typically 0.3 mm per sample.

## 12.4   Experimental Procedures

The multisensor data collection and reference measurement procedure involved several carefully determined steps. These are described below.

(1) **Preparation:**

  (a) Mount *Anthrobot-3* hand on the *PUMA* arm. Run the CALIBRATE command at the *VAL* prompt to calibrate the *PUMA* arm.

  (b) Position the hand in the camera FOV, with the palm facing upwards. Place object on the hand, and grasp it. Adjust the finger positions so that several tactile sensors are activated. The *Cyberglove* glove is used to teleoperate the hand, and make these adjustments. Lock the hand in the final grasp configuration.

  (c) Adjust external fixture, to be used for "freezing" the object pose when taking the reference measurements (Step 4). Practice this step, to make sure that the object can be clamped without disturbing the object, such that its weight is supported by the external fixture. Now, move this external fixture away from the camera FOV.

(2) **Tactile Data Collection:** The tactile sensors are placed at known locations on the finger-tips. Their locations w.r.t. the *PUMA* base frame are computed from the kinematics of the hand [164], given its location w.r.t. the *PUMA* base frame and corrected as described above. These contact points are saved into a file.

(3) **Vision Data Collection:** The background is adjusted to be black, to ensure that only the white object and the black hand holding it are visible in the camera FOV. The lighting is adjusted to minimize reflections, and uniformly illuminate the object. These experiments were carried out under ordinary fluorescent room lights directly overhead. For some experiments we had to use a light from below, to eliminate shadows and obtain more uniform illumination.

The POPEYE [23] software package is used to process the images, and extract the vertex features, which are saved into a file. The low-level vision preprocessing and feature extraction algorithms are controlled by several parameters, which must be adjusted for the

scene and the lighting conditions. Once these were appropriately
adjusted, they remained the same for the entire set of experiments.

(4) **Reference Measurement:**

   (a) The external fixture, which was prepared in Step 1c, is now
       slowly and carefully brought in, and tightened on the object.
       This must be accomplished without disturbing the object. The
       finger-tip positions are also marked on the object surface, so
       that they can be probed with a stylus later.

   (b) Now, open up the hand using the *Cyberglove* and move the
       arm and hand away. The object should not move, and is
       supported by the external fixture. Unmount the *Anthrobot-3*,
       and mount the stylus on the PUMA arm.

   (c) Probe several ($\geq$ 3) object vertices with the stylus, and save
       their coordinates in a file. These points are used to deter-
       mine the reference pose for evaluation of multisensor fusion
       estimate.

## 12.5   Experiments

As shown in the block diagram of the laboratory experiments (Figure 12.2),
the contact data features obtained from the hand finger-tips, the vertex
features obtained from the camera images, and the camera calibration are
inputs to the MOMFIS multisensor system, which implements the minimal representation size multisensor framework developed in this book. The
output of MOMFIS is an *estimated interpretation*, specified by the estimated object pose, and the correspondence assignments for the tactile and
vision data features. This estimated interpretation is compared against a
reference interpretation.

### 12.5.1   *Reference Interpretation*

The *reference pose* is computed by finding the least-squares-error (LSE)
fit of the model object vertices to the corresponding object vertex positions measured with a stylus attached to the PUMA arm (Step 4 in Section 12.4). The RMS error between the transformed model vertices and
their world coordinates is minimized to find the reference pose parameters.
It was convenient to use the same differential evolution [190] search algo-

rithm (see Section 8.5.3.2) used to perform the minimal representation size minimization, where the pose was represented as described in Section 11.1.1 on page 175 and the correspondence was known. The typical RMS error of a reference pose, obtained using 4-7 object vertices, was in the range of 2-12mm.

The *reference correspondences* for the tactile and vision data sets are assigned manually, by observation of the hand configuration and the camera images, respectively. The reference pose and the reference correspondences, specify a complete *reference interpretation* of the environment, for which we can compute the "reference" representation size.

### 12.5.2  *Interpretation Error*

The translation error between the estimated pose, and the reference pose (Figure 12.2 on page 205) is simply the Euclidean distance between the two values. The orientation error between the estimated and the reference pose is specified by the angle between the two rotation axes, and the difference in the rotational angle about them.

The correspondence error is specified by the number of mis-assignments in the estimated correspondence, w.r.t. the reference correspondence. This is reported in terms of **(a)** the number of modeled data features in the reference correspondence that were matched against a different model feature in the estimated correspondence, **(b)** the number of modeled data features in the reference correspondence that were labeled as outliers, and **(c)** the number of spurious data features (outliers) in the reference measurement that were modeled in the estimated correspondence.

The representation size deviation is the difference between the estimated interpretation representation size and the reference interpretation representation size.

### 12.5.3  *Results*

We conducted experiments with two different objects: *Lwedge* (Figure 11.7 on page 193) and *Pedestal* (Figure 11.8 on page 194), each in three different poses.

Encoded ellipsoidal accuracy and precision based uncertainty models (Section 6.3.1.1) were used for both the tactile and the vision sensors. For the vision sensor we used a two dimensional elliptical uncertainty region given by the pixel size of $8.3333 \times 10^{-03}$mm $\times$ $9.7166 \times 10^{-03}$mm, and a measurement space equal to the image size of $640 \times 480$ pixels. For the tactile sensor, we used an ellipsoidal ball of radius 0.1mm, equal to half the *PUMA* repeatability, and a measurement space of radius 150mm, equal to the *Anthrobot-3* workspace volume. These values are based on a *theoretical* justification of the accuracy and precision for the two sensors, and were kept the same for the entire set of experiments.

A one-to-one correspondence model was used for the vision sensor, while a many-to-one correspondence model was used for the tactile sensor. In both cases, a two-part fixed-length encoding algorithm (Section 7.3.2.2) was used to compute the sensor representation size, using the sensor models developed in Sections 11.2.1 and 11.2.2.

A differential evolution (DE) search algorithm (see Sections 8.5.2, 8.5.3.2, and 11.1.1) was used to search for the minimal representation size pose and the correspondence assignments for each data set. The translation search space was restricted to a 50x50x50 cube, in the interest of computation time. The entire space of orientations and correspondences was searched. All the experiments used a 100 member DE/best/2 differential evolution search algorithm (Section 8.5.2), with $F = 0.8$, and $CR = 0.8$.

The results of multisensor fusion using vision and touch data are tabulated in Table 12.5.3 on the facing page, and shown in Figures 12.4-12.21. Note that the camera calibration and vision preprocessing and feature extraction parameters were identical in Experiments 2-6. Table 12.5.3 summarizes the results using only the tactile data, while Table 12.5.3 summarizes the results using only the vision data. The CPU times reported in Tables 12.5.3, 12.5.3, and 12.5.3 are on a SPARC 20 Sun workstation.

Comparing the interpretation errors in Tables 12.5.3, 12.5.3, and 12.5.3, we find that multisensor fusion dramatically improved the pose estimates obtained from either sensor used independently. The run times are of the order of a few hundred seconds on a SPARC 20 computer.

Table 12.1 Summary of experimental results using multisensor vision (V) and touch (T) data to estimate pose and correspondences.

| Experiment Id | Sensor | M, Model Feature Count | N, Data Feature Count | Reference Outliers | Estimated Outliers | Modeled Mis-modeled | Modeled ← Unmodeled | Spurious → Modeled | Axis Error, degrees | Angle Error, degrees | Translation Error | Rep. Size Deviation | Ref. Pose RMS Error | Generations/CPU Seconds | Object Name/Figures |
|---|---|---|---|---|---|---|---|---|---|---|---|---|---|---|---|
| 1 | V | 18 | 16 | 8 | 4 | 0 | 0 | 4 | 0.06 | 2.20 | 1.9 | -73.1 | 1.7 | 339 | *Lwedge* 12.4,12.5,12.6 |
|   | T | 13 | 5 | 0 | 0 | 0 | 0 | 0 |      |       |      |       |      | 585 |      |
| 2 | V | 18 | 10 | 2 | 1 | 1 | 0 | 1 | 1.71 | 1.35 | 26.7 | -67.2 | 1.8 | 346 | *Lwedge* 12.7,12.8,12.9 |
|   | T | 13 | 9 | 0 | 0 | 0 | 0 | 0 |      |       |      |       |      | 569 |      |
| 3 | V | 18 | 7 | 1 | 0 | 1 | 0 | 1 | 0.13 | 2.70 | 8.81 | -41.2 | 1.2 | 321 | *Lwedge* 12.10,12.11,12.12 |
|   | T | 13 | 9 | 0 | 0 | 0 | 0 | 0 |      |       |      |       |      | 594 |      |
| 4 | V | 14 | 8 | 4 | 3 | 2 | 1 | 2 | 0.33 | -17.91 | 21.2 | -117.3 | 11.8 | 440 | *Pedestal* 12.13,12.14,12.15 |
|   | T | 9 | 9 | 0 | 0 | 6 | 0 | 0 |      |       |      |       |      | 596 |      |
| 5 | V | 14 | 10 | 4 | 2 | 2 | 0 | 2 | 0.06 | -6.67 | 12.9 | -104.0 | 11.6 | 407 | *Pedestal* 12.16,12.17,12.18 |
|   | T | 9 | 9 | 0 | 0 | 1 | 0 | 0 |      |       |      |       |      | 571 |      |
| 6 | V | 14 | 11 | 6 | 5 | 0 | 1 | 2 | 0.03 | 8.91 | 29.3 | -120.2 | 12.1 | 448 | *Pedestal* 12.19,12.20,12.21 |
|   | T | 9 | 9 | 0 | 0 | 1 | 0 | 0 |      |       |      |       |      | 599 |      |

Table 12.2   Summary of experimental results using only the tactile (T) data to estimate pose and correspondence.

| Experiment Id | Sensor | M, Model Feature Count | N, Data Feature Count | Reference Outliers | Estimated Outliers | Modeled Mis-modeled | Modeled → Unmodeled | Spurious → Modeled | Axis Error, degrees | Angle Error, degrees | Translation Error | Rep. Size Deviation | Ref. Pose RMS Error | Generations | Object Name |
|---|---|---|---|---|---|---|---|---|---|---|---|---|---|---|---|
| 1 | T | 13 | 5 | 0 | 0 | 3 | 0 | 0 | 1.01 | 70.07 | 7.7 | -47.7 | 1.7 | 385 | *Lwedge* |
| 2 | T | 13 | 9 | 0 | 0 | 7 | 0 | 0 | 0.90 | 21.09 | 13.4 | -56.8 | 1.8 | 210 | *Lwedge* |
| 3 | T | 13 | 9 | 0 | 0 | 9 | 0 | 0 | 0.49 | 115.51 | 30.4 | -45.9 | 1.2 | 239 | *Lwedge* |
| 4 | T | 9 | 9 | 0 | 0 | 6 | 0 | 0 | 1.40 | 10.47 | 32.6 | -110.6 | 11.8 | 370 | *Pedestal* |
| 5 | T | 9 | 9 | 0 | 0 | 6 | 0 | 0 | 0.05 | 13.38 | 20.7 | -85.1 | 11.6 | 376 | *Pedestal* |
| 6 | T | 9 | 9 | 0 | 0 | 7 | 0 | 0 | 1.08 | 5.58 | 28.9 | -92.1 | 12.1 | 267 | *Pedestal* |

Table 12.3   Summary of experimental results using only the vision (V) data to estimate pose and correspondence.

| Experiment Id | Sensor | $M$, Model Feature Count | $N$, Data Feature Count | Reference Outliers | Estimated Outliers | Modeled Mis-modeled | Modeled → Unmodeled | Spurious → Modeled | Axis Error, degrees | Angle Error, degrees | Translation Error | Rep. Size Deviation | Ref. Pose RMS Error | Generations | Object Name |
|---|---|---|---|---|---|---|---|---|---|---|---|---|---|---|---|
| 1 | V | 18 | 16 | 8 | 4 | 5 | 2 | 6 | 1.35 | 122.71 | 19.8 | -41.1 | 1.7 | 361 | Lwedge |
| 2 | V | 18 | 10 | 2 | 3 | 3 | 1 | 0 | 0.39 | 52.08 | 10.9 | -40.8 | 1.8 | 511 | Lwedge |
| 3 | V | 18 | 7 | 1 | 0 | 6 | 0 | 1 | 1.67 | 59.56 | 18.6 | -21.7 | 1.2 | 530 | Lwedge |
| 4 | V | 14 | 8 | 4 | 4 | 2 | 1 | 2 | 0.09 | 12.47 | 29.9 | -24.3 | 11.8 | 590 | Pedestal |
| 5 | V | 14 | 10 | 4 | 1 | 5 | 1 | 4 | 0.90 | 161.20 | 31.3 | -26.0 | 11.6 | 501 | Pedestal |
| 6 | V | 14 | 11 | 6 | 5 | 1 | 4 | 2 | 0.04 | 11.80 | 27.1 | -33.0 | 12.1 | 399 | Pedestal |

Fig. 12.4   Experiment 1 (*Lwedge*): Camera image and extracted vertex features.

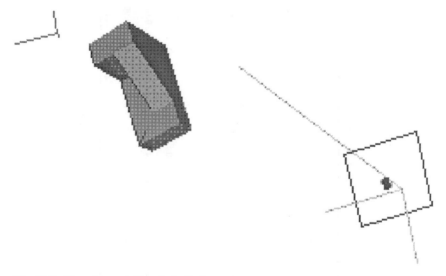

Fig. 12.5   Experiment 1 (*Lwedge*): Estimated interpretation relative to the camera and world reference frames.

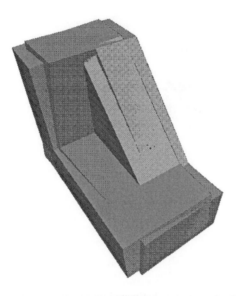

Fig. 12.6   Experiment 1 (*Lwedge*): Estimated and reference poses.

Fig. 12.7   Experiment 2 (*Lwedge*): Camera image and extracted vertex features.

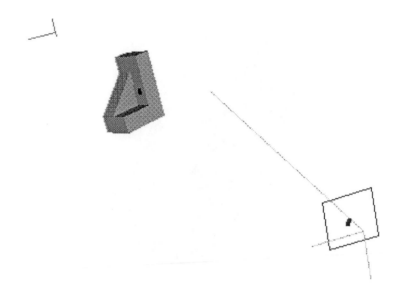

Fig. 12.8   Experiment 2 (*Lwedge*): Estimated interpretation relative to the camera and world reference frames.

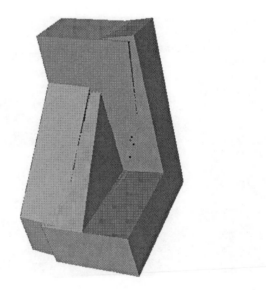

Fig. 12.9   Experiment 2 (*Lwedge*): Estimated and reference poses.

Fig. 12.10   Experiment 3 (*Lwedge*): Camera image and extracted vertex features.

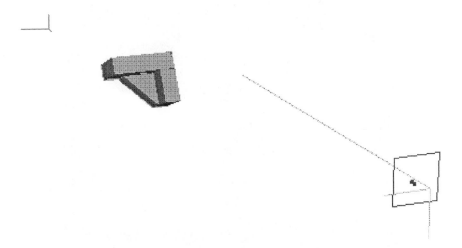

Fig. 12.11   Experiment 3 (*Lwedge*): Estimated interpretation relative to the camera and world reference frames.

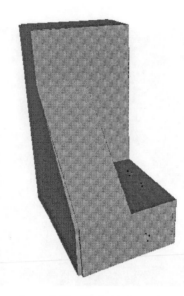

Fig. 12.12   Experiment 3 (*Lwedge*): Estimated and reference poses.

Fig. 12.13   Experiment 4 (*Pedestal*): Camera image and extracted vertex features.

Fig. 12.14   Experiment 4 (*Pedestal*): Estimated interpretation relative to the camera and world reference frames.

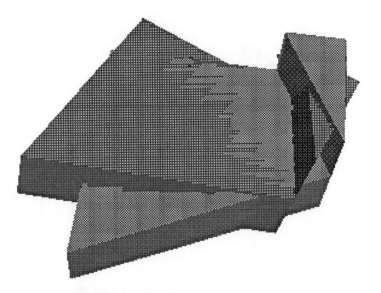

Fig. 12.15   Experiment 4 (*Pedestal*): Estimated and reference poses.

Fig. 12.16   Experiment 5 (*Pedestal*): Camera image and extracted vertex features.

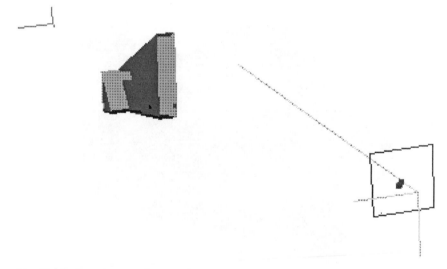

Fig. 12.17   Experiment 5 (*Pedestal*): Estimated interpretation relative to the camera and world reference frames.

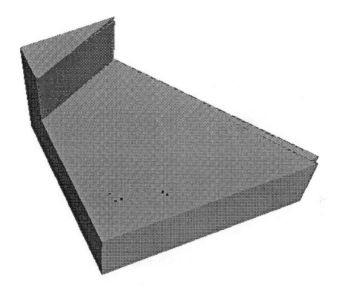

Fig. 12.18   Experiment 5 (*Pedestal*): Estimated and reference poses.

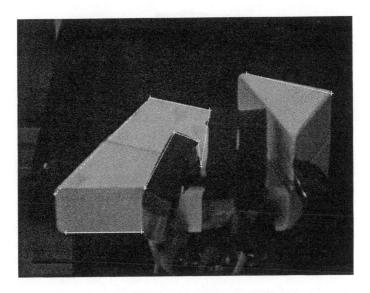

Fig. 12.19   Experiment 6 (*Pedestal*): Camera image and extracted vertex features.

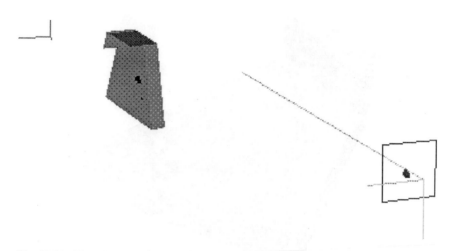

Fig. 12.20  Experiment 6 (*Pedestal*): Estimated interpretation relative to the camera and world reference frames.

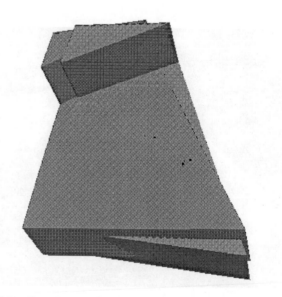

Fig. 12.21  Experiment 6 (*Pedestal*): Estimated and reference poses.

# Chapter 13

# Discussion of Experimental Results

In this chapter we discuss the experimental results in relation to the multisensor fusion and model selection framework developed in Part 2, and summarized in Chapter 9.

## 13.1 Multisensor Fusion and Model Selection

The experimental results described in Section 12.5.3, demonstrate the automatic selection of environment model class (object identity), the environment parameter values (object pose), and the sensor data correspondences (touch and vision correspondences), in the minimal representation size multisensor fusion and model selection framework.

### 13.1.1 *Environment Model Parameterization and Data Scaling*

From Table 12.5.3 on page 213 and Figures 12.6-12.21, it can be seen that the estimated pose parameters produced by minimal representation size multisensor fusion of tactile and vision data, was close to the reference pose for nearly all the experiments.

Experiment 4 had the largest errors, both in the estimated pose and the vision and touch correspondence. This is attributed to the vision data being lined up on the upper half of the object (Figure 12.13 on page 220), with 4/8 spurious data features. Also, in this experiment, the tactile data has large errors w.r.t. the reference pose. Upon examining the estimated pose (Figure 12.15 on page 221), we find that for this particular experimen-

225

tal geometry, it is a "reasonable" explanation of the observed vision and tactile data. With better lighting (or a lower image processing threshold), this result could be improved. However, for the purpose of these experiments, it was more important to illustrate a typical range of results without optimal tuning of conditions and vision preprocessing parameters for each experiment.

The multisensor fusion pose parameter estimate dramatically improved the estimates obtained from either sensor used independently (Tables 12.5.3 and 12.5.3). The interpretation error when using only a single sensor, was much larger for all the experiments. Closer examination of the data reveals some interesting observations:

- **Incomplete Data:** The tactile data is incomplete since at most three different surfaces are contacted by the fingers, in these experiments. Several different poses of the object may result in the same contact configuration—therefore multiple solutions exist. Indeed, successive searches produced vastly different solutions for the nearly the same estimated representation size values in Experiments 3 and 5, when using only the touch data.

- **Spurious Data:** The vision data set contains spurious data features introduced by the presence of the hand holding the object, as can be seen from Table 12.5.3 on page 213 and Figures 12.4, 12.7, 12.10, 12.13, 12.16, 12.19. In Experiments 1, 4, and 6, at least 50% the vision data features were spurious. These spurious data features compete with the "good" data features, for a minimal representation size match, and can produce incorrect solutions, when using only the vision data.

In these experiments, multisensor fusion of tactile and vision sensors compensates for the deficiencies in using either sensor alone, and results in significantly smaller . interpretation errors. The minimal representation size multisensor and model selection framework automatically selected an appropriate combination of the vision and touch sensor constraints to determine the pose parameters. Within these constraints, the resulting pose estimation errors are comparable in magnitude to the sensor errors themselves, and constitute a successful practical demonstration of the minimal representation size multisensor fusion framework for this application.

## 13.1.2 *Data Subsample Selection*

The vision and touch data correspondence errors were small in all the experiments (Table 12.5.3). In those cases, where there was a mis-assignment, the selected correspondence was still meaningful for that particular spatial arrangement of data features and selected pose parameters.

In the present implementation, hidden object vertices are not removed (for the sake of computational efficiency), resulting in an artificially larger number of vision model features. In Experiments 1, 4, and 6, at least half the vision data features were spurious; the spurious data features got matched to these "extra" model features, which would otherwise be occluded, thus producing a somewhat larger vision correspondence error. This phenomenon is more severe when using the vision data alone (Table 12.5.3). However using the touch data compensates for this effect, and results in the selection of vision and touch correspondences with significantly smaller errors (Table 12.5.3) than when either sensor is used alone.

The correspondence selection/data subsampling property of the framework is illustrated by a plot of minimal representation size vs. the number of outliers, in Figure 13.1. This plot was generated from the Experiment 1 multisensor data (Section 12.5.3), by controlling the number of vision outliers as follows:

(1) Subsample the set of vision data features by eliminating the data features with the largest representation size (for the "best" interpretation reported in Table 12.5.3 on page 213).

(2) Find the minimal representation size interpretation using this smaller set of vision data features and the original set of tactile data features.

(3) Add the representation size of the eliminated vision data features (treating them as outliers), to this minimal representation size. This gives the total representation size of the original multisensor data.

(4) Plot this total representation size against the total number of vision outliers in this interpretation (number of eliminated data features + number of outliers found in the subsampled vision data set) in this interpretation.

The minimum of this plot corresponds to the underlying interpretation which best explains the observed data, and is reported in Table 12.5.3. The

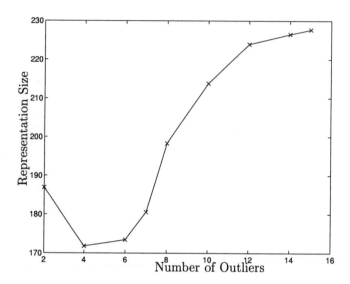

Fig. 13.1   Representation size vs. the number of vision outliers for Experiment 1 multi-sensor data. The minimal representation size interpretation with 4 outliers, corresponds to the underlying model itself, and gives the best tradeoff between the number of un-modeled features (model size) and the encoded error residuals.

minimal representation size solution trades-off between the number of un-modeled data features and the modeled error residuals. This is essentially a tradeoff between the model size (number of bits needed to represent the un-modeled data features) and encoded error residuals (number of bits needed to encode the errors with respect to the model features). As the number of unmodeled data features is "forced" to be larger, the total representation size increases, since these data features must now be explicitly encoded in the measurement space of the sensor. As the number of outliers is "forced" to be smaller, the total representation size increases again, since these data features must now be encoded w.r.t. some model features with large error residuals.

The minimal representation size criterion selects the simplest correspondence model, which best explains the observed data (Section 7.4.3). Given, two environment model parameter instantiations (for the same model structure, at the same level of parameter resolution), the correspondence model is chosen to minimize the data representation size—the criterion trades-off between the error residuals and the model size (measurement space encoding of unmodeled data features) within the constraints of the corre-spondence model. This tradeoff between model size and error residuals is analogous to the one seen in the polynomial model order selection problem, discussed in Section 1.3.2 (Figure 1.4 on page 11).

### 13.1.3 *Environment Model Class Selection*

In the laboratory experiments, only the model class (i.e., the object iden-tity) must be selected, since the number of (pose) parameters is the same for all model structures (object shapes) in the library, and we assume that the pose parameters are estimated upto some fixed parameter resolution. The library contains two objects: *Lwedge* and *Pedestal* (see Figures 11.7 and 11.8). The *Pedestal* object is "simpler" than the *Lwedge* object, since it has fewer number of vertices and faces.

The model class selection property of the minimal representation size framework (Section 7.4.1) is shown in Table 13.1. Three different corre-spondence encoding algorithms, $\mathcal{A}$, are used to illustrate model class se-lection. For each sensor, a two-part encoding scheme is used, which may employ either **(a)** a *fixed*-length correspondence encoding algorithm (see Equations 7.25 and 7.26 on page 106), or **(b)** an *entropy* based correspon-dence encoding algorithm (see Equations 7.19 and 7.20 on page 104). The results using different combinations of these coding algorithms are shown in Table 13.1. The framework selected the correct model class for all the experiments, when using the fixed length encoding algorithm for touch, and either encoding algorithm for the vision sensor. Except for Experiment 2, the others show a separation of $\sim 10$ bits or more for the two object shapes, and are also correctly classified using an entropy based encoding algorithm for touch data.

Since, in these experiments $N \sim M$ for both the vision (V) and the touch (T) sensors, using the fixed length coding algorithm gives smaller codelengths for the many-to-one touch correspondence, while using the en-tropy based coding algorithm algorithm gives smaller codelengths for the one-to-one vision correspondence. This is confirmed in Table 13.1, where using the fixed length encoding for the touch and the entropy based encod-ing for the vision (Equations 7.23 and 7.24 on page 106) give the smallest representation sizes.

For Experiment 2, the minimal representation size using either shape is comparable. Upon closer examination we find that, for this data set, the

Table 13.1   Model class selection for experimental data, relative to a library containing two shapes: *Lwedge* and *Pedestal*. The shape which gives the smallest representation size is selected, as indicated by the boldface letters.

| Expt# | Lwedge | Pedestal | Coding Algorithm, $\mathcal{A}$ |
|---|---|---|---|
| 1 | **195.07** | 216.80 | V, T: fixed |
| *Lwedge* | **202.21** | 204.76 | V, T: entropy |
| | **189.61** | 197.64 | V: entropy, T: fixed |
| 2 | **157.49** | 159.59 | V, T: fixed |
| *Lwedge* | 165.38 | **160.37** | V, T: entropy |
| | **153.47** | 154.64 | V: entropy, T: fixed |
| 3 | **126.48** | 144.99 | V, T: fixed |
| *Lwedge* | **133.48** | 139.41 | V, T: entropy |
| | **121.57** | 133.68 | V: entropy, T: fixed |
| 4 | 161.81 | **145.54** | V, T: fixed |
| *Pedestal* | 167.61 | **149.30** | V, T: entropy |
| | 151.53 | **139.41** | V: entropy, T: fixed |
| 5 | 206.36 | **161.46** | V, T: fixed |
| *Pedestal* | 210.62 | **168.31** | V, T: entropy |
| | 196.80 | **158.42** | V: entropy, T: fixed |
| 6 | 196.11 | **177.24** | V, T: fixed |
| *Pedestal* | 207.27 | **175.91** | V, T: entropy |
| | 191.20 | **168.27** | V: entropy, T: fixed |

error residuals using either shape are comparable—therefore either shape may constitute a "reasonable" explanation of the observed data. Although using a fixed-length encoding for touch results in correct classification, this is no longer the case when using an entropy based (sub-optimal) encoding algorithm—the simpler model (*Pedestal*) is selected as an explanation of the observed data.

These results illustrate the tradeoff between the environment model class representation and the data representation size. Given, two environment model structures (with the same number of parameters at same level of parameter resolution), the model class is chosen to minimize the total representation size—the criterion trades-off between the data representa-

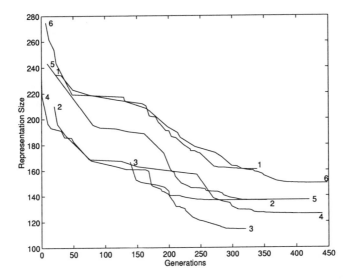

Fig. 13.2 Representation size versus the number of generations for Experiments 1-6, using a 100-member DE/best/2, $F = 0.8, CR = 0.8$.

tion size and the model representation size. When the data representation size for two environment models are comparable, the simpler model (with few model features) is selected.

## 13.2 Multisensor Fusion Search Algorithms

### 13.2.1 *Differential Evolution Search*

Figure 13.2 shows the plots of representation size against the number of generations, for the differential evolution search (Section 8.5.3.2) on the six laboratory experiments. As can be seen from these plots, the DE fully converged by 400 generations in all cases. The run-times are of the order of a few minutes on a SPARC 20 (Table 12.5.3). For practical use of the method in "real-time" control applications, this global search procedure should be augmented with an incremental search procedure, which sequentially updates the estimated model as more data is acquired during the execution of the real-time control task.

Figure 13.3 on page 233 shows the progress of the differential evolution algorithm on Experiment 1 multisensor data. The algorithm starts with a population of 100 randomly chosen individuals. The population cluster shrinks around the minimal representation size interpretation, as can be seen from Figure 13.3.

Figure 13.4 shows a plot of the percentage convergence against the number of generations for the DE. As can be seen from this plot the orientation parameters converge first followed by the translational pose parameters. This is typical of the DE, for this application.

When the observed data is *incomplete* or inconsistent, multiple interpretations result in nearly the same representation size. In such cases, successive runs of the DE may produce varying results, due the multiple global minima.

As the noise in the observed data increases, or as the number of observed data features increases, the DE converges correctly more often, and is more reliable. This is perhaps due to the minimal representation size search landscape becoming smoother with deeper valleys, when the observation errors or the number of data features increase.

### 13.2.2 *Comparison with a Binary Genetic Algorithm*

We compared the performance of DE to that of a simple binary genetic algorithm (GA), which uses the one point crossover and mutation operators for reproduction, and stochastic universal sampling for natural selection [66, 136]. The GA follows the same outline as the DE, described in Section 8.5.3.2. The environment model parameter vector is represented as a *binary string*, with 32 bits per element and arranged linearly, as shown in Figure 11.2. A population of 100 individuals was used, and the crossover probability was 0.5, while the mutation probability was chosen to be 0.05. These GA parameters were selected after some trial and error, to minimize the representation size at the end of the search.

The performance of this GA is compared with the 100-member DE/best/2 algorithm, with F = 0.8, CR = 0.8, on Experiment 1 multisensor data. Both the GA and the DE were run until either, the population reached 99% convergence, or until the maximum time limit of 900 CPU seconds is exceeded, or until a maximum of 1000 generations is exceeded. The parameter tolerance for the DE convergence was set to 0.01. The convergence was checked every 10 generations.

Figure 13.5 shows a plot of the representation size against the number

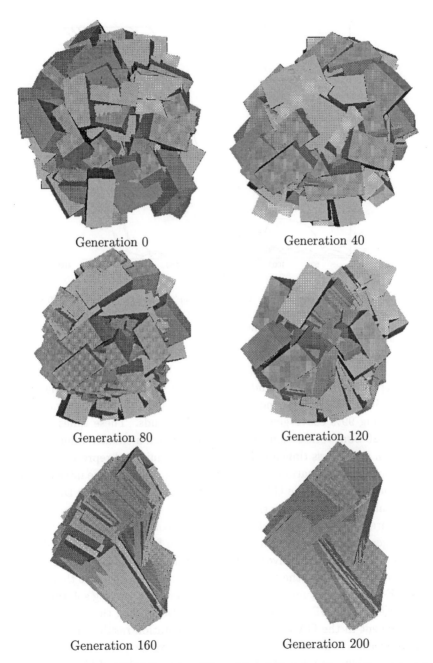

Generation 0          Generation 40

Generation 80          Generation 120

Generation 160          Generation 200

Fig. 13.3  Progress of the 100-member differential evolution algorithm on Experiment 1 multisensor data. DE/best/2, $F = 0.8, CR = 0.8$.

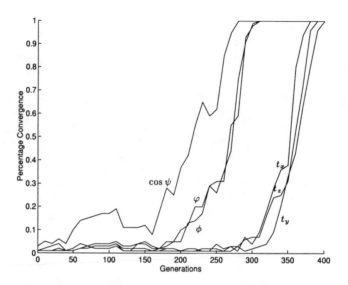

Fig. 13.4   Percentage convergence vs. the number of generations for the 100-member DE/best/2, F $= 0.8,$ CR $= 0.8$, on Experiment 1 multisensor data. The DE ran until the population reached 99% convergence. The parameter tolerance for convergence was set to 0.01, and checked every 10 generations.

of generations for the DE and the GA. The GA ran until the maximum time limit of 900 CPU seconds was exceeded, while the DE terminated up-on reaching 99% convergence in 634 CPU seconds. These times are on a SPARC 20 Sun workstation. The DE found much a smaller value of representation size in less time and fewer number of minimal representation size correspondence evaluations (Section 8.3.1) than the GA. The interpretation error for the DE and the GA are tabulated in Table 13.2. The DE solution had smaller interpretation errors.

Figure 13.6 shows a plot of the percentage convergence against the number of generations for the GA. As can be seen from this plot, it is difficult to discern any regular convergence pattern for the pose parameters.

In our experience, the GA convergence was less consistent than that of the DE—successive runs of the GA produced widely varying solutions when compared to the DE, on the same problem. Decreasing the number of bits per element in the GA binary string representation resulted in premature convergence. This can be explained by the coarser parameter resolution, due to the fewer number of bits. Increasing the number of bits per param-

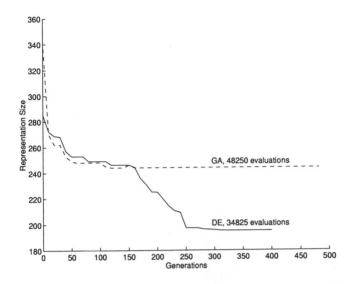

Fig. 13.5 Representation size vs. the number of generations for the 100-member DE/best/2, F = 0.8, CR = 0.8, and the 100-member Binary GA, with crossover probability 0.5 and mutation probability 0.05, on Experiment 1 multisensor data.

Table 13.2 Comparison of the interpretation error using a DE and a GA for Experiment 1 multisensor data.

| Algorithm | DE | GA |
|---|---|---|
| Rotation axis error | 0.05605° | 0.16949° |
| Rotation angle error | 2.2022° | 4.0203° |
| Translation error | 1.867 | 27.092 |
| Vision correspondence error | 4 | 8 |
| Touch correspondence error | 0 | 0 |
| Representation size deviation | -73.102 | -24.064 |
| CPU Time (seconds) | 634 | > 900 |

eter resulted in very long computation times with very slow convergence, if at all. This can be explained by the higher dimensionality of the GA binary search space, in which the more significant bits of the parameter vector are treated the same as the less significant bits.

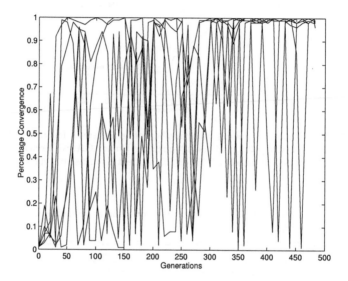

Fig. 13.6   Percentage convergence vs. the number of generations for the 100-member Binary GA, with crossover probability 0.5 and mutation probability 0.05, on Experiment 1 multisensor data. The GA ran until the the time limit of 900 CPU seconds expired. The convergence was checked every 10 generations.

# Chapter 14

# Conclusion

## 14.1 Summary

As applications become more complex, the selection of the model, parameters, and data samples are more difficult to determine a priori, and therefore the availability of a consistent framework across a wide variety of possible model choices and problem domains is important. In this book, we described a new approach to multisensor fusion and model selection which uses **(a)** generalized constraint manifolds constraint manifold to describe the relation between observed data features and environment model features, **(b)** encoded accuracy and precision or probability distributions to describe the sensor observation errors and the measurement space, and **(c)** a minimal representation size criterion which serves as a *universal sensor independent yardstick* for fusing multisensor data.

This leads to a unified framework for multisensor fusion and model selection characterized by **(i)** applicability to a wide variety of problem domains, **(ii)** consistent scaling of disparate sensor data on the basis of sensor accuracy and sensor resolution, and **(iii)** automatic selection of model class and number of parameters, the parameter values, parameter resolution, and the sensor data correspondences.

### 14.1.1 *Multisensor Fusion*

With respect to the multisensor fusion problems and issues discussed in Section 3.1.2, the minimal representation size multisensor fusion framework **(1)** requires calibrated sensors to achieve sensor registration, **(2)** naturally sup-

237

ports the fusion of homogeneous and heterogeneous sensors since disparate sensor data is expressed in the number of bits of information (representation size) which is sensor independent, **(3)** scales the sensor data according to sensor accuracy and resolution, and automatically rejects outliers in each data set, **(4)** effectively fuses incomplete and spurious sensor data, provided the sensors are consistent, i.e., observe the same environment, **(5)** automatically selects the correspondence between the observed data and environment model features, **(6)** assumes that sensors are independent, **(7)** can be used for sensor data at various granularity levels, provided one can compute the representation size, **(8)** assumes that the environment is static, i.e., all the data has been gathered, **(9)** does not directly address dynamic updates of the model estimates as more data is collected, **(10)** has a modular structure in which additional sensors contribute an additional term to the total multisensor representation size (Figure 7.3).

### 14.1.2  *Model Selection*

The multisensor fusion framework provides a consistent basis for selection among alternative statistical models, different sensor combinations, different parameterization, scaling, and subsampling choices. When probabilistic uncertainty models or approximations are difficult to determine, direct encoding of the data features in terms of sensor accuracy and data precision provides a consistent alternative. Often observed data features and model features from disparate sensors will be unrelated quantities. While conventional methods of sensor fusion must often make arbitrary weighing among such data, the minimal representation size criterion provides a natural framework for fusion of different quantities, using the representation size as a universal yardstick to combine disparate sensor data. In this framework, the more accurate sensors provide more information, and their contribution is inherently weighted more highly. In practice, a major advantage of this approach is the attainment of consistent results without the introduction of problem specific heuristics or arbitrary weight factors.

The "model selection" happens at many different levels: **(1)** correspondence model selection (data subsample selection), given an environment model class, number of parameters, parameter values and their resolution, **(2)** environment model parameter value and resolution selection (model parameterization), given an environment model class and number of parameters, **(3)** environment model order selection (number of parameters),

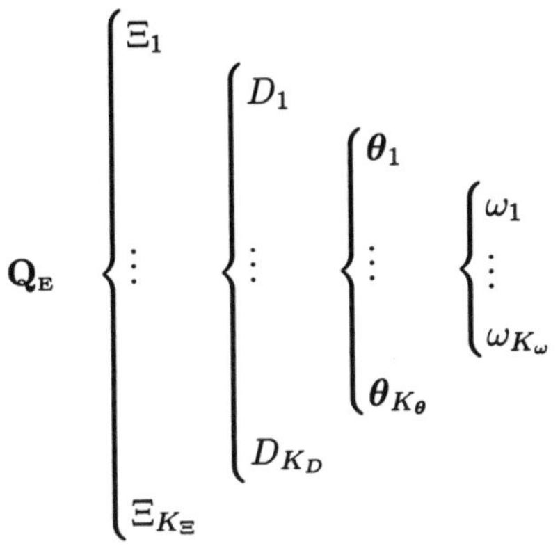

$$\mathbf{Q_E} \left\{ \begin{array}{l} \Xi_1 \\ \vdots \\ \Xi_{K_\Xi} \end{array} \left\{ \begin{array}{l} D_1 \\ \vdots \\ D_{K_D} \end{array} \left\{ \begin{array}{l} \boldsymbol{\theta}_1 \\ \vdots \\ \boldsymbol{\theta}_{K_\theta} \end{array} \left\{ \begin{array}{l} \omega_1 \\ \vdots \\ \omega_{K_\omega} \end{array} \right. \right. \right. \right.$$

Fig. 14.1 Model selection hierarchy. Given a model library $\mathbf{Q_E}$, one may choose from a number of model structures $\Xi$. Given a model structure, one may choose from a number of different model-orders $D$. Given a model-order, one may chose from number of parameter vectors $\boldsymbol{\theta}$. Given a parameter vector, one may choose from a number of correspondence assignments $\omega$.

given an environment model class, and (4) environment model class selection, given a library of environment model classes. The model library $\mathbf{Q_E}$ may be regarded as being composed of a hierarchy of models different levels of abstraction, as shown in Figure 14.1. At each level of this hierarchy, the minimal representation size framework trades-off between the "model representation size" and the "data representation size", selecting the simplest model that best explains the data. The "data representation size" at each level is comprised of the model + data representation size at the next successive level. This model selection behavior at various levels is illustrated in Figure 13.1 for correspondence selection, Table 12.5.3 for model parameterization, Figure 1.3 for model order selection, and Table 13.1 for model class selection.

This framework for model selection may also be used in conjunction with standard statistical techniques for parameter estimation and fusion. A suitable error residual encoding method (e.g., the commonly used Gaussian pdfs as in Section 6.3.2 which result in least squares error estimation, or an encoded accuracy and precision based uncertainty model as in Section 6.3.1) may be used for parameter estimation within the framework. The framework can automatically select the correspondences, the environment model class, and its parameters, at an appropriate parameter resolution (dependent on the observed data, the sensor uncertainties, and the sensor constraints). The minimum complexity choice provided by the minimal representation size criterion is intuitively coupled to the concept of the efficient utilization of sensor information and the optimal choice of sensors and 'data features, for a given task.

### 14.1.3    *Search Algorithms*

We described generic algorithm templates for finding for a minimal representation size interpretation. The hypothesize and test algorithms are exhaustive, but may take a long time to complete. The evolution programs are converge much faster; however they do not guarantee finding a minimal representation size interpretation.

The search algorithms can be organized according to the model hierarchy shown in Figure 14.1. The minimal representation size correspondence is selected for each environment parameter value (Section 8.3). The minimal representation size parameter value and resolution are selected for each model class (Sections 8.4 and 8.5). The minimal representation size model class is selected from the environment model library, by sequentially iterating through the list of environment model classes.

As developed here, the search problem is difficult because we have posed it as a broad search over many general parameter configurations (local minima), and have chosen not to impose heuristic constraints to simplify the search. In practice, there are many such heuristics which may be imposed for specific problems, and they would often improve the execution time of the search. However, in this study the focus has been on the nature of the representation itself and the associated search algorithms, rather than on building a practical system. The understanding and performance of these measures and algorithms provides the basis for further improvement in practical systems.

### 14.1.4    *Object Recognition*

We applied the multisensor fusion and model selection technique to object recognition and pose estimation in two and three dimensions, using vision, touch, and grasp sensors. In the laboratory experiments, a robot hand holds an object in front of a camera, and the identity and pose of the object is automatically determined by fusing the vision and touch data, *while the hand is holding the object.* This is a very *basic sensing capability,* required by intelligent robotic systems for accomplishing precise manipulation tasks.

The laboratory experiments demonstrate the automatic selection of environment model class (object identity), the environment parameter values (object pose), and the sensor data correspondences (touch and vision correspondences), in the minimal representation size framework for multisensor fusion and model selection.

The laboratory experiments confirm the need for using multiple sensors in this application. The tactile sensors provide incomplete data, and therefore a unique interpretation cannot be found. The vision sensors contain spurious data features due to the presence of the hand in the image, which often results in an incorrect interpretation. By fusing data from touch and vision sensors, the correct interpretation can be consistently found.

The differential evolution search algorithm was used to solve these problems, and its time complexity grows linearly with the number of data features from each sensor. Several independent populations may be evolved in parallel, to improve the reliability of the DE from 81% (Section 11.5.1) to, say, 99%. The run-times are a few hundred seconds on a SPARC 20.

In practice, robotic manipulation such as described here must be done in real-time, and it is clear that the general evolutionary algorithms are too slow to estimate pose at sampling speeds for continuous motion. In a real-time system, one would use the generalized algorithm described here to initialize and occasionally update the pose estimate, while a much faster real-time update algorithm based, for example, on a Kalman filter would provide real-time feedback.

## 14.2 Research Topics

### 14.2.1 *Multisensor Fusion*

#### 14.2.1.1 *Model Parameter Resolution and Sensor Selection*

As discussed in Section 7.4.1, the minimal representation size criterion can be used to automatically select the "optimal" *parameter resolution* for given set of observed multisensor data. We can analyze the parameter estimation algorithm to derive an expression for this parameter resolution (Section 6.4) as a function of the sensor accuracies and the number of data features from each sensor. This analysis would provide bounds on the parameter resolution, and can be used to define the number of bits per parameter in the model representation size (Equation 7.12 on page 98). As discussed in Section 7.4.1, these bounds are related to the *Fisher information*, and the resulting model size increases as the Fisher information increases. More analysis is required to formally develop these relationships. It seems there is a richer underlying mathematical structure relating the sensor accuracies to the optimal parameter resolution according to the sensor constraints and number of data features.

The expression for parameter resolution may also be used to solve the inverse problem of sensor design, where given a desired parameter resolution (Section 6.4), we want to find:

- Which sensors (i.e., which data features) should be used?
- What is the accuracy desired of those sensors?
- What is the minimum number of data features which must be acquired from each sensor?

This would allow us to design optimal task-specific sensor configurations, from an information-theoretic standpoint.

### 14.2.1.2  *Sensor Data Acquisition Planning*

The multisensor fusion framework assumes that the multisensor data has been acquired by some "appropriate" means. Given a multisensor configuration, the data acquisition strategy itself can be the subject of further study. Following is a list of some of the issues:

- How to choose the sensor parameters (placement etc.) to minimize the sensing cost (number of data features, number of sensing operations, etc.)?
- How to select the next sensing operation? The notion of information gain, introduced in Section 7.5.1, may be useful as a criterion for choosing the "next-operation", as the one which maximizes the information gain.
- When is the sensor data incomplete? How to automatically acquire additional data features, to make the data set complete?

## 14.2.1.3 *Dependent Sensors*

We assumed that the sensors are independent (Section 7.2.1). Exploiting the dependencies between multiple sensors should result in smaller multisensor data representation size, and a more efficient utilization of codewords. Since the environment provides the link between multiple sensors, one approach for exploiting sensor dependency is to *reconstruct* an intermediate environment representation from the sensor data and compute its residuals w.r.t. the environment model. The reconstruction algorithm arguments must also be encoded in the expression for the total multisensor representation size. Exploiting dependencies between sensors may lead to improved model estimates.

## 14.2.2  *Computational Efficiency*

### 14.2.2.1  *Sequential Interpretation Estimation Algorithms*

The framework utilizes all the acquired sensor data, to perform multisensor fusion and model selection. While this *batch processing* is useful for initializing the model estimate, a natural extension of this work is to develop techniques for *sequentially updating* the model estimate as more data is acquired.

Dynamic incremental updating of the model estimate, as more data is collected, is very important from a computational efficiency and real-time implementation standpoint (see Section 14.1.4). In a real-time system, one could use the global search algorithm using all the acquired data to initialize and periodically update the model estimate, while a much faster sequential update algorithm using just the newly acquired data, would provide real-time model estimates.

The *information gain* measure (Section 7.5.1) may be used as a criterion for *sequential estimation*: as additional data is acquired, the estimated interpretation is updated (say, by a perturbation around the current best interpretation) to maximize the information gain. This may result in a local search which could be much faster than the global search needed to find the best interpretation using all the data.

### 14.2.2.2 *Batch Interpretation Estimation Algorithms*

Improved *batch estimation* algorithms can be developed, to search for the best interpretation using all of the acquired data.

In the differential evolution search algorithm described in Section 8.5.3.2, each individual in the population represents a parameter vector (pose), and the minimal representation size correspondence is computed during the fitness evaluation of an individual. A separate DE search is run for each environment model structure (object id). The minimal representation size model among all these searches is used as the selected interpretation. As the size of the environment model library grows, this approach may quickly become impractical. It may be advantageous to encode the model structure (object id) and the correspondence as additional elements in the representation of an individual, and let the DE search for the model structures, parameters, and correspondences simultaneously. Additional domain specific knowledge may be useful in developing hybrid evolution programs to efficiently find the best interpretation.

In Section 8.5.2.6 we presented an informal analysis of the differential evolution search algorithm which relates the choice of search parameters to the population diversity and selection pressure. These should be formalized, and the mechanics of DE analyzed to develop a theoretical foundation for this search method. This understanding may be useful in developing more reliable search algorithms.

The generic multisensor fusion algorithms presented in Chapter 8 can be parallelized.

The differential evolution algorithm (Section 8.5.3.2) can be parallelized by independently evolving multiple populations (each population may be an independent thread of execution), and keeping the best individual. Our experience suggests that, we should run several DE's in parallel, and perhaps devise a periodic *migration* strategy between the multiple evolving populations, to further improve the overall search performance.

For the hypothesize and test algorithm (Section 8.4), the CDFS index combinations can be enumerated lexicographically, and partitioned into several mutually exclusive index sets, one per independent thread of execution. The threads operate in parallel, and the final solution is obtained by taking a minimum of the solutions found by each of the independent threads.

Each thread of execution may run on a separate processor, or may be distributed across a network of computers [93, 94].

### 14.2.2.3  *Domain Heuristics*

As noted in Chapter 8, there may be several domain heuristics which may be used to reduced the search space, and quickly compute a minimal representation size model estimate. In a practical application, domain knowledge and heuristics should be exploited to develop efficient algorithms for searching the minimal representation size model.

## 14.2.3  *Multisensor Object Recognition*

### 14.2.3.1  *Articulated Shapes and Additional Sensors*

In the multisensor object recognition applications presented (Chapters 9.2 and 10.5.2), we assumed that the only unknown parameters were the pose of the object. Thus, the number of model parameters was the same for all environment models, and only the model class had to be selected. As a next step, *articulated* shapes requiring additional articulation parameters could be used. The environment model library would include unarticulated and articulated shapes with varying number of parameters.

Based on the observed multisensor data, the correct shape class, and the number of articulated parameters can be chosen. In addition the values of the articulated parameters, pose parameters, and the data correspondences can be automatically determined by minimal representation size multisensor fusion framework.

The laboratory experiments used a single camera and the touch contact data. It is straightforward to incorporate *additional cameras* in the minimal representation size multisensor fusion criterion of Equation (7.10), using the vision sensor model described in Section 11.2.1. Each additional camera would have to be individually calibrated. It is also straightforward to incorporate the grasp data, if available, using the grasp sensor model in Section 11.2.3.

Other sensors (i.e., additional types of data features) may be used (and may be perhaps necessary), to deal with complex environments containing multiple objects, severe occlusions, and distractor objects.

### 14.2.3.2  *Demonstration of a Robotic Assembly Task*

In the laboratory experiments on multisensor object recognition and pose estimation, the inputs are the vision and tactile data features, and the outputs are the object identity, pose, and the vision and tactile data correspondences. This output is an essential *prerequisite* for implementing hand regrasping strategies and executing motion plans, to accomplish precise manipulation tasks.

Consider the assembly task, in which the hand picks an object part from a parts-bin and holds it in front of a camera to estimate object identity, pose, and the vision and touch correspondences. These estimates are then used as inputs to grasp planners and motion planners, which determine a regrasping strategy and a motion plan, to place the object in a desired goal configuration, and build a complex assembly. Additional intermediate sensing operations may be used to provide feedback during task execution. A practical implementation might require development of *hybrid algorithms*, which exploit knowledge of the domain, action commands, and other constraints in conjunction with the minimal representation size framework, to efficiently compute the pose estimates.

A complete demonstration of such a "pick-and-place" assembly task, would be an important step towards building autonomous intelligent robotic systems.

### 14.2.4  *New Applications*

The methodology presented in this book is a general approach to fusion which is not restricted to geometric pose estimation, and in fact may be applied to a variety of problems in model identification from a wide perspective, including model-based identification of parameters from noisy data and prioritization in noisy data sets. The extension of this approach to many different types of models and estimation problems [16, 114, 210, 213, 215] is an important element of this methodology and its use for general parametric model estimation with noisy data is open for further research.

# Appendix A

# List of Symbols

*Minimal Representation Size*

| | |
|---|---|
| $\mathcal{D}$ | data |
| $\mathbf{q}$ | model |
| $\mathbf{Q}$ | model library |
| $\mathcal{A}$ | encoding algorithm |
| $\mathcal{U}$ | universal turing machine |
| $\mathcal{S}(\cdot)$ | representation |
| $\mathcal{L}[\cdot]$ | representation size |
| $\mathcal{G}[\mathcal{D}^+\|\mathcal{D}]$ | information gain due to additional data |
| $\mathcal{C}(\cdot)$ | relative confidence |
| $H(\cdot)$ | entropy |
| $\mathcal{J}[\cdot]$ | redundancy |
| $\mathcal{I}(n)$ | representation size of a non-negative integer $n$ |

*Environment and Sensor Models*

| | |
|---|---|
| $\boldsymbol{\theta}$ | environment model parameters |
| $\boldsymbol{\Theta}$ | set of legitimate environment model parameter vectors |
| $D$ | dimensionality of the parameter vector |
| $\Xi$ | environment model structure |
| $\mathbf{q}_{\mathrm{E}}$ | environment model, $\Xi(\boldsymbol{\theta})$ |
| $\mathbf{Q}_{\mathrm{E}}$ | library of environment models |
| $L$ | number of environment model structures in the library |
| $\Psi$ | sensor uncertainty model describing observation errors |
| $\Upsilon$ | sensor measurement space |

| | |
|---|---|
| $\boldsymbol{\alpha}$ | sensor calibration |
| $\mathbf{q}_\text{s}$ | sensor coefficients, $(\Psi, \Upsilon, \boldsymbol{\alpha})$ |
| $S$ | number of sensors |
| | |
| $\mathbf{z}$ | a data feature |
| $n$ | dimensionality of the data features |
| $\mathbf{Z}$ | set of observed data features |
| $N$ | number of data features |
| $\tilde{N}$ | number of modeled data features |
| | |
| $\mathbf{y}$ | a model feature |
| $\mathbf{Y}$ | set of extracted environment model features |
| $M$ | number of model features |
| | |
| $\omega$ | correspondence mapping data features to model features |
| $\Omega$ | set of legitimate correspondence mappings |
| $f$ | relative frequency vector |
| $\mathbf{f}$ | set of legitimate relative frequency vectors |
| | |
| $\mathbf{h}$ | constraint relating data and model features, $\mathbf{h}(\mathbf{y}; \mathbf{z}) = 0$ |
| $K$ | number of constraint equations |
| $\text{DCM}(\mathbf{y})$ | data constraint manifold (DCM) |
| $\tilde{n}$ | dimensionality of the DCM |
| | |
| $\mathcal{F}$ | environment model feature extractor, $\{\mathbf{y}\} = \mathcal{F}(\mathbf{q}_\text{E}; \boldsymbol{\alpha})$ |
| $\mathcal{H}$ | sensing channel, $(\mathcal{F}, \mathbf{h}, \omega)$ |
| | |
| $\mathbf{q}$ | multisensor observation model, $(\mathbf{q}_\text{E}, \{\mathbf{q}_\text{s}, \mathcal{H}\})$ |
| $\Lambda$ | sensor accuracy |
| $\lambda$ | sensor precision |
| $\rho(\boldsymbol{\theta})$ | sensor resolution |

*Evolution Programs*

| | |
|---|---|
| $P$ | number of parents |
| $R$ | number of offspring |
| $\mathcal{P}(\cdot)$ | a member of the parent population |
| $\mathcal{P}'(\cdot)$ | a member of the intermediate population |

$\mathcal{P}''(\cdot)$      a member of the offspring population

**p**      the search problem

$\Phi(\cdot)$      fitness function

$\nu_c$      crossover operator parameters

$\nu_m$      mutation operator parameters

$\nu_s$      selection operator parameters

*Differential Evolution*

F      differential scaling factor

CR      crossover probability

$\gamma$      greediness

$K$      number of differentials

$\mathcal{D}$      indices of the parameter vector elements to be perturbed

*Object Recognition and Pose Estimation*

$\mathcal{R}[\cdot]$      rotation operator

$\mathcal{T}[\cdot]$      pose transformation operator

$\vec{t}$      translation vector

$q$      quaternion describing a rotation

$R$      homogeneous $3 \times 3$ rotation matrix

$\mathrm{ROT}(\hat{n}, \phi)$      rotation about and axis $\hat{n}$ by an angle $\phi$

$s$      scale

$\vec{p}$      shape vertex

**e**      shape edge joining vertex $\vec{p}^{\,a}$ to vertex $\vec{p}^{\,b}$

**f**      polygonal object face, whose plane is specified by three vertices $\vec{p}^{\,a}$, $\vec{p}^{\,b}$, $\vec{p}^{\,c}$, or by an outward surface normal $\vec{m}$ and perpendicular distance $c$ from the origin

$\vec{r}$      observed data vertex

$\vec{w}$      observed data contact point

**s**      observed data face, whose plane is specified by an outward surface normal $\vec{n}$ and perpendicular distance $d$ from the origin

$\vec{t}_c$      position of the world frame relative to the camera frame

| | |
|---|---|
| $R_c$ | orientation of the world frame relative to the camera frame |
| $f$ | focal length |
| $\kappa$ | radial length distortion coefficient |
| $C_u, C_v$ | lens center |
| $K_{ccd}$ | number of CCD elements in the camera |
| $K_U$ | number of pixels in a scan-line |
| $K_V$ | number of scan-lines |
| $s_{rate}$ | sampling rate mismatch |
| $\sigma_u, \sigma_v$ | pixel size |
| | |
| $\zeta, \xi$ | intermediate parametric variables |

# Appendix B

# Object-Oriented Software Architecture

The **MO**del-based **M**ultisensor **F**usion and **I**nterpretation **S**ystem (MOM-FIS) is a highly reusable *object-oriented application framework*, implemented in the *C++* programming language [191]. MOMFIS takes a modular component based approach to software development. The MOMFIS software application framework classes embody the mathematical multisensor fusion framework described in this book. To apply the MOMFIS application framework on a new problem domain, the programmer has to derive a few classes and "fill in the blanks", for that specific problem domain.

Figure B.1 shows the MOMFIS application framework class diagram, and its specialization to two and three dimensional object recognition problem domains, using the notation described in [63]. The classes participating in this framework are described below.

MomfisEncoder A *class utility* [21] which provides methods to encode primitive types (Section 5.5). Used by many other classes in the framework (e.g., *MomfisModel, MomfisSensor, MomfisUncertainty, MomfisCorresp*, and their derived classes).

MomfisFactory A *singleton* factory class, which uses a variation of the *prototype* design pattern [63] to create concrete objects of the following types: *MomfisUncertainty, MomfisCorresp, MomfisSearch*, and *MomfisDomain*. Each *concrete* subclass of these classes must register a creation method with the MomfisFactory object.

MomfisTool Multisensor data fusion and model selection tool with an interactive textual user interface. Allows the user to choose an application domain, load experimental data, generate simulated data, run various search algorithms, print performance statistics, and

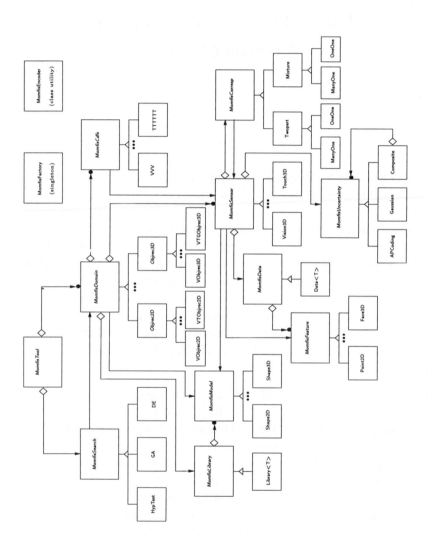

Fig. B.1  Class diagram of the MOMFIS application framework. The abstract classes are shown in slanted type face.

view/display the data and results.

*MomfisSearch* Abstract base class for multisensor fusion search algorithms (Chapter 8). Finds the minimal representation size interpretation (environment model and multisensor correspondences) for a given a problem domain object.

> MomfisHypTest A generic hypothesize and test algorithm (Section 8.4).
>
> MomfisGA A generic binary genetic search algorithm [66] (Section 8.5).
>
> MomfisDE A generic differential evolution search algorithm (Section 8.5.3.2).

*MomfisDomain* Abstract base class for multisensor fusion problem domains. Has an environment model library, multiple sensor models, and CDFS definitions. Stores a pointer to the selected environment model in the library.

> *Objrec2D* Two-dimensional multisensor object recognition problem domain (Chapter 10).
>
> > VObjrec2D Two-dimensional object recognition using only vision data.
> >
> > TObjrec2D Two-dimensional object recognition using only touch data.
> >
> > VTObjrec2D Two-dimensional object recognition using vision and touch data.
>
> *Objrec3D* Three-dimensional multisensor object recognition problem domain (Chapter 11).
>
> > VObjrec3D Three-dimensional object recognition using only vision data.
> >
> > TObjrec3D Three-dimensional object recognition using only touch data.
> >
> > GObjrec3D Three-dimensional object recognition using only grasp data.
> >
> > VTObjrec3D Three-dimensional object recognition using vision and touch data.
> >
> > VTGObjrec3D Three-dimensional object recognition using vision, touch, and grasp data.

*MomfisCdfs* Abstract base class for specifying and solving CDFS (Section 8.2.2).

> VV2D Two-dimensional VV CDFS (Section F.1).
> VTT2D Two-dimensional VTT CDFS (Section F.2).
> TTTT2D Two-dimensional TTTT CDFS (Section F.3).
> VVV Three-dimensional VVV CDFS (Section I.1).
> VVTT Three-dimensional VVTT CDFS (Section I.2).
> VTTTT Three-dimensional VTTTT CDFS (Section I.3).
> GGG Three-dimensional GGG CDFS (Appendix G).

*MomfisLibrary* An abstract base class for a library of environment models.

> Library<T> A generic environment model library template class, parameterized by the *type* of the environment model class.

*MomfisModel* An abstract base class for environment models. Provides methods for computing representation size, and randomly instantiating the environment model parameters.

> Shape2D Two-dimensional object shape models (Section 10.1).
> Shape3D Three-dimensional object shape models (Section 11.1).

*MomfisSensor* An abstract base class for sensor models. Keeps a pointer to an environment model in the library. Has a correspondence model, an uncertainty model, and the raw sensor data. Provides methods for extracting model features, conditioning raw data features, and computing representation size.

> Vision2D Two-dimensional vision sensor model (Section 10.2.1).
> Touch2D Two-dimensional touch sensor model (Section 10.2.2).
> Vision3D Three-dimensional vision sensor model (Section 11.2.1).
> Touch3D Three-dimensional touch sensor model (Section 11.2.2).
> Grasp3D Three-dimensional grasp sensor model (Section 11.2.3).

*MomfisCorresp* An abstract base class for correspondence models (Section 6.2). Provides methods for finding the minimal representation size correspondence, given an instantiated environment model (Section 8.3).

> MomfisTwopart Two-part discrete correspondence (Section 7.3.2.2). Implements a known correspondence (Table 7.1). Derived

classes implement the search algorithms described in Section 8.3.1. Uses fixed-length correspondence encoding (Section 7.3.2.2).

MomfisTwopartManyOne Discrete two-part many-to-one encoding.

MomfisTwopartManyOne Discrete two-part one-to-one encoding.

MomfisMixture Mixture continuous correspondence (Section 7.3.2.3). Implements a known correspondence (Table 7.1). Derived classes implement the search algorithms described in Section 8.3.2.

MomfisMixtureManyOne Mixture relative frequency many-to-one encoding.

MomfisMixtureManyOne Mixture relative frequency one-to-one encoding.

*MomfisUncertainty* Abstract base class for $n$-dimensional sensor uncertainty models. Provides methods for computing representation size, and generating simulated noise vectors and outlier data features.

MomfisAPCoding Encoded accuracy and precision based uncertainty model (Section 6.3.1).

MomfisGaussian A Gaussian probabilistic uncertainty model (Section 6.3.2).

MomfisComposite A composite uncertainty model, using the *composite* design pattern [63]. Splits an $n$-dimensional vector sequence into several component sequences, each described by an uncertainty model of type *MomfisUncertainty*. This is useful when the data feature vector is composed of several unrelated feature spaces, as is the case for the grasp data features (Section 11.2.3).

*MomfisData* Abstract base class for sensor data.

Data<T> A generic sensor data template class, parameterized by the *type* of the data features.

*MomfisFeature* Abstract base class for data and model features.

Point2D A point in two-dimensional space.

Line2D  A finite line segment in two-dimensional space.
Point3D  A point in three-dimensional space.
Plane3D  A plane in three-dimensional space.

Face3D  A finite planar face in three-dimensional space.

# Appendix C

# Error Residuals

The error residual, $\Delta\mathbf{z}$, in Equation (7.15) can be approximated using the first order Taylor's series expansion of the constraint equation, $\mathbf{h}(\mathbf{y};\mathbf{z}) = 0$, about the point $\mathbf{z} = \tilde{\mathbf{z}} + \Delta\mathbf{z}$; thus

$$\mathbf{h}(\mathbf{y};\mathbf{z}) = \mathbf{h}(\mathbf{y};\tilde{\mathbf{z}} + \Delta\mathbf{z})$$

$$\approx \mathbf{h}(\mathbf{y};\tilde{\mathbf{z}}) + \left[\frac{\partial\mathbf{h}(\mathbf{y};\mathbf{z})}{\partial\mathbf{z}}\right]_{\mathbf{z}=\tilde{\mathbf{z}}}\Delta\mathbf{z} \qquad\qquad (C.1)$$

$$= \left[\frac{\partial\mathbf{h}(\mathbf{y};\mathbf{z})}{\partial\mathbf{z}}\right]_{\mathbf{z}=\tilde{\mathbf{z}}}\Delta\mathbf{z},$$

since $\mathbf{h}(\mathbf{y};\tilde{\mathbf{z}}) = 0$ by the definition of DCM. Let $J(\tilde{\mathbf{z}})$ be the $K \times n$ *Jacobian matrix* of partial derivatives of $\mathbf{h}$ w.r.t. $\mathbf{z}$, evaluated at $\tilde{\mathbf{z}}$ ($K$ constraints in $n$-dimensional measurement space)

$$J(\tilde{\mathbf{z}}) \triangleq \left[\frac{\partial\mathbf{h}(\mathbf{y};\mathbf{z})}{\partial\mathbf{z}}\right]_{\mathbf{z}=\tilde{\mathbf{z}}}$$

Substituting in Equation (C.1), the error $\Delta\mathbf{z}$ is approximately,

$$\Delta\mathbf{z} \approx [J(\tilde{\mathbf{z}})^{\mathrm{T}}J(\tilde{\mathbf{z}})]^{-1}J(\tilde{\mathbf{z}})^{\mathrm{T}}\mathbf{h}(\mathbf{y};\mathbf{z}). \qquad\qquad (C.2)$$

Since $\tilde{\mathbf{z}}$ is unknown, we approximate

$$J(\tilde{\mathbf{z}}) = J(\mathbf{z}) \qquad\qquad (C.3)$$

which holds for small $\Delta\mathbf{z}$, as $\Delta\mathbf{z} \to 0$, and is a reasonable approximation when the tangent plane $J(\mathbf{z})$ varies slowly with $\mathbf{z}$. Therefore, for small $\Delta\mathbf{z}$

$$\Delta\mathbf{z} \approx [J(\mathbf{z})^{\mathrm{T}}J(\mathbf{z})]^{-1}J(\mathbf{z})^{\mathrm{T}}\mathbf{h}(\mathbf{y};\mathbf{z}). \tag{C.4}$$

which can be evaluated given an observed data feature $\mathbf{z}$. Once $\Delta\mathbf{z}$ is approximated using Equation (C.4), we can calculate $\tilde{\mathbf{z}} = \mathbf{z} - \Delta\mathbf{z}$, and use it in Equation (C.2) to get an improved estimate of $\Delta\mathbf{z}$. This can be repeated in an *iterative* procedure to compute a reasonably accurate value of $\Delta\mathbf{z}$. This method will find a point $\tilde{\mathbf{z}}$ on the DCM surface "close" to $\mathbf{z}$, but does not guarantee that $\tilde{\mathbf{z}}$ will satisfy the *boundary conditions* associated with the constraint equation.

Equations (C.1) and (C.3), hold *exactly* when the constraint equation $\mathbf{h}(\mathbf{y};\mathbf{z})$ is a *linear* function of the data feature $\mathbf{z}$. In that case $J(\mathbf{z}) = J$, a constant, and we get

$$\Delta\mathbf{z} = [J^{\mathrm{T}}J]^{-1}J^{\mathrm{T}}\mathbf{h}(\mathbf{y};\mathbf{z}).$$

i.e. Equations (C.2) and (C.4) hold exactly for linear constraints.

## Appendix D

# Mixture Distributions

Consider a mixture of $M + 1$ random variables, $X_j, j = 0, \ldots, M$, "mixed" according to an independent discrete $M + 1$ valued random variable $W$, as shown in Figure D.1. The resulting mixture random variable $X = X_W$ has a distribution [151, 197]

$$\mathrm{p}_X(x) = \sum_{j=0}^{M} \Pr\{W = j\} \cdot \mathrm{p}_{X_j}(x),$$

where components $X_j$ may be discrete or continuous valued random variables.

What is the representation size of $X = x$? In a *two-part* encoding scheme, we would encode the value of $W = j$ in $-\log_2 \Pr\{W = j\}$ bits, and encode the value $x$ according to the distribution $\mathrm{p}_{X_j}(x)$ in $-\log_2 \mathrm{p}_{X_j}(x)$

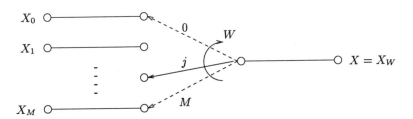

Fig. D.1 The random variable $X$ is a mixture of $M + 1$ random variables $X_0, \ldots, X_M$, chosen according to an $M + 1$ valued discrete random variable $W$.

bits; thus

$$
\begin{aligned}
\mathcal{L}[X = x] &= \mathcal{L}[W = j] + \mathcal{L}[X = x \,|\, W = j] \\
&= -\log_2 \Pr\{W = j\} - \log_2 \mathrm{p}_{X_j}(x) \quad bits. \qquad \text{(D.1)}
\end{aligned}
$$

Note that $\mathrm{p}_{X,W}(x,j) = \mathrm{p}_{X_j}(x) \cdot \Pr\{W = j\}$, and therefore, $\mathcal{L}[X = x] = -\log_2 \mathrm{p}_{X,W}(x,j)$, since $W$ is *independent* of $X_0, \ldots, X_M$ random variables*. In a *mixture* encoding scheme, we would encode the mixture random variable $X$ directly using the mixture distribution

$$
\mathcal{L}[X = x] = -\log_2 \mathrm{p}_X(x) = -\log_2 \sum_{j=0}^{M} \Pr\{W = j\} \cdot \mathrm{p}_{X_j}(x) \quad bits. \qquad \text{(D.2)}
$$

The mixture encoding gives a shorter encoding, which can be proved as follows.

**Theorem D.1**   *Given a mixture random variable $X$ of $M+1$ components $X_0, \ldots, X_M$, chosen according to a discrete $M+1$ valued random variable $W$, the mixture representation size of an observation $X = x$ is less than or equal to the two-part representation size. Furthermore, the* average *mixture representation size of $X$ is given by the entropy $H(X)$, and is less than or equal to the* average *two-part representation size of $X$ given by the joint entropy*

$$
H(X, W) = H(W) + H(X \,|\, W) = H(X) + H(W \,|\, X).
$$

**Proof.**   Observe that $\forall j = 0, \ldots, M$

$$
\Pr\{W = j\} \cdot \mathrm{p}_{X_j}(x) \leq \sum_{j=0}^{M} \Pr\{W = j\} \cdot \mathrm{p}_{X_j}(x) \qquad \text{(D.3)}
$$

since all the terms are non-negative. The two-part representation size expression of Equation (D.1) may be written as

$$
\mathcal{L}[X = x] = -\log_2 \Pr\{W = j\} \cdot \mathrm{p}_{X_j}(x)
$$

Comparing it with the mixture representation size expression of Equation (D.2), and noting that $-\log x$ *is a monotonically decreasing function*

---

*$X_0, \ldots, X_M$ may be be independent

*of its argument* $x$, we get the desired result, viz. $\forall j = 0, \ldots, M$

$$-\log_2 \Pr\{W = j\} \cdot p_{X_j}(x) \geq -\log_2 \sum_{j=0}^{M} \Pr\{W = j\} \cdot p_{X_j}(x),$$

i.e. the mixture representation size of $X = x$ is less than or equal to the two-part representation size.

Taking the expectation w.r.t. $p_{X,W}(x,j)$ on both sides of the above expression, and noting that $p_{X,W}(x,j) = p_{X_j}(x) \cdot \Pr\{W = j\}$, we get

$$H(X, W) \geq H(X)$$

where $H(X, W)$ is the average two-part representation size or the joint entropy of $X$ and $W$, and $H(X)$ is the average mixture representation size or entropy of $X$. Also from Cover and Thomas [40], $H(X, W) = H(W) + H(X \mid W) = H(X) + H(W \mid X)$. $\qquad\square$

In other words, mixture encoding gives us shorter encodings most of the time. The savings come from the fact that an observation $X = x$ can be explained by several overlapping distributions for which $p_{X_j}(x) \neq 0$. Rather than assigning a separate codeword per component distribution, as in two-part encoding, we share a single codeword for those values of $x$ for which the component distributions overlap.

**Corollary D.1** *The mixture and two-part representation sizes are identical if and only if the component distributions are non-overlapping for any* $X = x$, *i.e.*

$$\forall x \quad p_{X_j}(x) \neq 0 \quad \text{for exactly one} \quad j \in \{0, \ldots, M\}.$$

*Proof.* If the component distributions are non-overlapping, the summation on the right hand side of Inequality (D.3) reduces to a single term, and thus it holds with equality $\forall x$. Therefore the two-part and mixture representation sizes are identical.

If the mixture and two-part representation are identical, the Inequality (D.3) must hold with equality $\forall x$. Since all the terms in Inequality (D.3) are non-negative, and w.l.o.g. $\Pr\{W = j\} \neq 0, \forall j$, therefore $p_{X_j}(x) = 0$ for all but *one* $j$ on the right hand side. $\qquad\square$

# Appendix E

# Properties of Mixture Representation Size

## E.1 Convexity

The mixture representation size expression of Equation (7.28) is *strictly convex* $\cup$ in $\mathbf{f}$. Rewriting $\mathcal{L}[\omega, \mathbf{z} \mid \mathcal{A}, \mathbf{q}_E, \mathbf{Q}_E] = \mathcal{L}[\omega, \mathbf{z} \mid \mathbf{Y}] \equiv \mathcal{L}[\mathbf{f}, \mathbf{z} \mid \mathbf{Y}]$, we have $\forall \lambda \in [0, 1]$,

$$\mathcal{L}[\lambda \mathbf{f}^a + (1 - \lambda)\mathbf{f}^b, \mathbf{z} \mid \mathbf{Y}]$$

$$= -\log_2(\sum_{j=0}^{M} [\lambda f_j^a + (1 - \lambda)f_j^b] \cdot 2^{-\mathcal{L}[\mathbf{z} \mid \mathbf{y}_j]})$$

$$= -\log_2(\lambda \cdot \sum_{j=0}^{M} f_j^a \cdot 2^{-\mathcal{L}[\mathbf{z} \mid \mathbf{y}_j]} + (1 - \lambda) \cdot \sum_{j=0}^{M} f_j^b \cdot 2^{-\mathcal{L}[\mathbf{z} \mid \mathbf{y}_j]})$$

$$< -\lambda \cdot \log_2(\sum_{j=0}^{M} f_j^a \cdot 2^{-\mathcal{L}[\mathbf{z} \mid \mathbf{y}_j]}) - (1 - \lambda) \cdot \log_2(\sum_{j=0}^{M} f_j^b \cdot 2^{-\mathcal{L}[\mathbf{z} \mid \mathbf{y}_j]})$$

$$= \lambda \cdot \mathcal{L}[\mathbf{f}^a, \mathbf{z} \mid \mathbf{Y}] + (1 - \lambda) \cdot \mathcal{L}[\mathbf{f}^b, \mathbf{z} \mid \mathbf{Y}]$$

since $\log(\cdot)$ is a strictly convex $\cap$ function.

In other words, a *unique global minima* exists in the space of all possible $(M + 1)$-dimensional vectors. Note that $\sum_{j=0}^{M} f_j = 1$ must be satisfied. A *global minima* exists in the space of all $(M + 1)$-dimensional frequency distributions, but may not be unique due to this constraint.

## E.2　Upper Bound

The representation size size of a modeled data feature is bounded above by
the unmodeled representation size, $\mathcal{L}[\mathbf{z} \mid \mathbf{y}_0]$ (Equation 7.18 on page 103).

Using this fact, we can establish an upper bound on the mixture representation size as follows.

$$\mathcal{L}[\omega, \mathbf{z} \mid \mathbf{Y}] = -\log_2(f_0 \cdot 2^{-\mathcal{L}[\mathbf{z} \mid \mathbf{y}_0]} + \sum_{j=1}^{M} f_j \cdot 2^{-\mathcal{L}[\mathbf{z} \mid \mathbf{y}_j]})$$

$$= -\log_2(f_0 \cdot 2^{-\mathcal{L}[\mathbf{z} \mid \mathbf{y}_0]} + (1 - f_0) \cdot \sum_{j=1}^{M} \frac{f_j}{\sum_{j=1}^{M} f_j} \cdot 2^{-\mathcal{L}[\mathbf{z} \mid \mathbf{y}_j]})$$

$$\overset{(a)}{\leq} -\log_2(f_0 \cdot 2^{-\mathcal{L}[\mathbf{z} \mid \mathbf{y}_0]})$$

$$= -\log_2(f_0 \cdot 2^{-\mathcal{L}[\mathbf{z} \mid \mathbf{y}_0]})$$

$$= -\log_2 f_0 + \mathcal{L}[\mathbf{z} \mid \mathbf{y}_0],$$

since $\sum_{j=1}^{M}(f_j / \sum_{j=1}^{M} f_j) \cdot 2^{-\mathcal{L}[\mathbf{z} \mid \mathbf{y}_j]}$ is a mixture probability distribution
of $M$ components, and therefore lies in the interval $[0, 1]$. In step (a) we
replaced it by the infimum value of 0. The inequality follows from the fact
that $-\log_2(\cdot)$ is monotonically decreasing.

## E.3　Lower Bound

A lower bound for the mixture representation size can be established similarly.

$$\mathcal{L}[\omega, \mathbf{z} \mid \mathbf{Y}] = -\log_2(f_0 \cdot 2^{-\mathcal{L}[\mathbf{z} \mid \mathbf{y}_0]} + \sum_{j=1}^{M} f_j \cdot 2^{-\mathcal{L}[\mathbf{z} \mid \mathbf{y}_j]})$$

$$= -\log_2(f_0 \cdot 2^{-\mathcal{L}[\mathbf{z} \mid \mathbf{y}_0]} + (1 - f_0) \cdot \sum_{j=1}^{M} \frac{f_j}{\sum_{j=1}^{M} f_j} \cdot 2^{-\mathcal{L}[\mathbf{z} \mid \mathbf{y}_j]})$$

$$\overset{(a)}{\geq} -\log_2(f_0 \cdot 2^{-\mathcal{L}[\mathbf{z} \mid \mathbf{y}_0]} + (1 - f_0))$$

$$= -\log_2(f_0 \cdot 2^{-\mathcal{L}[\mathbf{z} \mid \mathbf{y}_0]} + 1 - f_0)),$$

since $\sum_{j=1}^{M}(f_j/\sum_{j=1}^{M} f_j) \cdot 2^{-\mathcal{L}[z\,|\,y_j]}$ is a mixture probability distribution of $M$ components, and therefore lies in the interval $[0,1]$. In step (a) we replaced it by the supremum value of 1. Once again, the inequality follows from the fact that $-\log_2(\cdot)$ is monotonically decreasing.

# Appendix F

# Computing Two-Dimensional Pose from Vision and Touch Data

## F.1  VV

The VV CDFS aligns two arbitrary vision data features $\vec{r}_{i_1}$, $\vec{r}_{i_2}$ with two arbitrary vision model features $\mathcal{T}[\vec{p}_{j_1}]$, $\mathcal{T}[\vec{p}_{j_2}]$ respectively:

$$\vec{r}_{i_1} = \mathcal{T}[\vec{p}_{j_1}] = se^{\imath\phi_z}\vec{p}_{j_1} + \vec{t} \tag{F.1}$$

$$\vec{r}_{i_2} = \mathcal{T}[\vec{p}_{j_2}] = se^{\imath\phi_z}\vec{p}_{j_2} + \vec{t} \tag{F.2}$$

This results in 4 real equations in 4 unknowns (the four pose parameters), and can be solved to find $\mathcal{T}[\cdot]$.

The VV constraint equations are solved as follows. Subtracting Equation (F.2) from Equation (F.1),

$$\vec{r}_{i_1} - \vec{r}_{i_2} = se^{\imath\phi_z}(\vec{p}_{j_1} - \vec{p}_{j_2})$$

$$\Rightarrow se^{\imath\phi_z} = \frac{\vec{r}_{i_1} - \vec{r}_{i_2}}{\vec{p}_{j_1} - \vec{p}_{j_2}} \tag{F.3}$$

The scale, $s$, and rotation, $\phi_z$, are given by the amplitude and phase of $se^{\imath\phi_z}$ respectively:

$$s = \frac{\|\vec{r}_{i_1} - \vec{r}_{i_2}\|}{\|\vec{p}_{j_1} - \vec{p}_{j_2}\|} \tag{F.4}$$

$$e^{\imath\phi_z} = \frac{1}{s} \cdot \frac{\vec{r}_{i_1} - \vec{r}_{i_1}}{\vec{p}_{j_1} - \vec{p}_{j_2}} = \frac{\vec{r}_{i_1} - \vec{r}_{i_2}}{\|\vec{r}_{i_1} - \vec{r}_{i_2}\|} \cdot \frac{\|\mathcal{T}[\vec{p}_{j_1}] - \mathcal{T}[\vec{p}_{j_2}]\|}{\vec{p}_{j_1} - \vec{p}_{j_2}} \tag{F.5}$$

Substituting $se^{\imath\phi_z}$ from Equation (F.3) into Equation (F.1), we get the

267

translation,

$$\vec{t} = \vec{r}_{i_1} - se^{i\phi_z}\vec{p}_{j_1}$$

$$= \vec{r}_{i_1} - \frac{\vec{r}_{i_1} - \vec{r}_{i_2}}{\vec{p}_{j_1} - \vec{p}_{j_2}} \cdot \vec{p}_{j_1}$$

$$= \frac{\vec{r}_{i_2} \cdot \vec{p}_{j_1} - \vec{r}_{i_1} \cdot \vec{p}_{j_2}}{\vec{p}_{j_1} - \vec{p}_{j_2}} \tag{F.6}$$

## F.2   VTT

The VTT CDFS aligns an arbitrary vision data feature $\vec{r}_{i_1}$ and two arbitrary touch data features $\vec{w}_{i_2}$, $\vec{w}_{i_3}$ with an arbitrary vision model feature $\mathcal{T}[\vec{p}_{j_1}]$ and two arbitrary touch model features $\mathcal{T}[\mathbf{e}_{j_2}]$, $\mathcal{T}[\mathbf{e}_{j_3}]$ respectively:

$$\vec{r}_{i_1} = se^{i\phi_z}\vec{p}_{j_1} + \vec{t} \tag{F.7}$$

$$\vec{w}_{i_2} = se^{i\phi_z}\{\zeta_{j_2}\vec{p}^{\,b}_{j_2} + (1 - \zeta_{j_2})\vec{p}^{\,a}_{j_2}\} + \vec{t} \tag{F.8}$$

$$\vec{w}_{i_3} = se^{i\phi_z}\{\zeta_{j_3}\vec{p}^{\,b}_{j_3} + (1 - \zeta_{j_3})\vec{p}^{\,a}_{j_3}\} + \vec{t} \tag{F.9}$$

This results in 6 real equations in 6 unknowns (the four pose parameters, and two $\zeta$'s), and can be solved to find $\mathcal{T}[\cdot]$.

The VTT constraint equations are solved as follows. Subtracting Equation (F.7) from Equations (F.8) and (F.9) we get

$$\vec{w}_{i_2} - \vec{r}_{i_1} = se^{i\phi_z}\{\zeta_{j_2}(\vec{p}^{\,b}_{j_2} - \vec{p}^{\,a}_{j_2}) + \vec{p}^{\,a}_{j_2} - \vec{p}_{j_1}\}$$

$$\vec{w}_{i_3} - \vec{r}_{i_1} = se^{i\phi_z}\{\zeta_{j_3}(\vec{p}^{\,b}_{j_3} - \vec{p}^{\,a}_{j_3}) + \vec{p}^{\,a}_{j_3} - \vec{p}_{j_1}\}$$

$$\Rightarrow \quad se^{i\phi_z} = \frac{\vec{w}_{i_2} - \vec{r}_{i_1}}{\{\zeta_{j_2}(\vec{p}^{\,b}_{j_2} - \vec{p}^{\,a}_{j_2}) + \vec{p}^{\,a}_{j_2} - \vec{p}_{j_1}\}}$$

$$= \frac{\vec{w}_{i_3} - \vec{r}_{i_1}}{\{\zeta_{j_3}(\vec{p}^{\,b}_{j_3} - \vec{p}^{\,a}_{j_3}) + \vec{p}^{\,a}_{j_3} - \vec{p}_{j_1}\}} \tag{F.10}$$

Equation (F.10) gives two real equations in the two real unknowns $\zeta_{j_2}$ and $\zeta_{j_3}$. They form a system of *linear equations*, as can be seen by rearranging Equation (F.10):

$$\mathrm{Re}[(\vec{w}_{i_2} - \vec{r}_{i_1}) \cdot \{\zeta_{j_3}(\vec{p}^{\,b}_{j_3} - \vec{p}^{\,a}_{j_3}) + \vec{p}^{\,a}_{j_3} - \vec{p}_{j_1}\}]$$

$$= \mathrm{Re}[(\vec{w}_{i_3} - \vec{r}_{i_1}) \cdot \{\zeta_{j_2}(\vec{p}^{\,b}_{j_2} - \vec{p}^{\,a}_{j_2}) + \vec{p}^{\,a}_{j_2} - \vec{p}_{j_1}\}]$$

$$\mathrm{Im}[(\vec{w}_{i_2} - \vec{r}_{i_1}) \cdot \{\zeta_{j_3}(\vec{p}^{\,b}_{\ j_3} - \vec{p}^{\,a}_{\ j_3}) + \vec{p}^{\,a}_{\ j_3} - \vec{p}_{j_1}\}]$$
$$= \mathrm{Im}[(\vec{w}_{i_3} - \vec{r}_{i_1}) \cdot \{\zeta_{j_2}(\vec{p}^{\,b}_{\ j_2} - \vec{p}^{\,a}_{\ j_2}) + \vec{p}^{\,a}_{\ j_2} - \vec{p}_{j_1}\}]$$

and can be solved for $\zeta_{j_2}$ and $\zeta_{j_3}$ using Gaussian elimination. Once $\zeta_{j_2}$ and $\zeta_{j_3}$ are found, they are substituted into Equation (F.10) to compute $se^{i\phi_z}$, giving the scale $s$ and rotation $\phi_z$. Substituting $se^{i\phi_z}$ back into Equation (F.7), we get the translation $\vec{t} = \vec{r}_{i_1} - se^{i\phi_z}\vec{p}_{j_1}$. Note that if $\zeta_{j_2}, \zeta_{j_3} \notin [0, 1]$, the VTT combination is rejected and another one chosen.

## F.3  TTTT

The TTTT CDFS aligns four arbitrary touch data features $\vec{w}_{i_1}$, $\vec{w}_{i_2}$, $\vec{w}_{i_3}$, $\vec{w}_{i_4}$ with four arbitrary touch model features $\mathcal{T}[\mathbf{e}_{j_1}]$, $\mathcal{T}[\mathbf{e}_{j_2}]$, $\mathcal{T}[\mathbf{e}_{j_3}]$, $\mathcal{T}[\mathbf{e}_{j_4}]$ respectively:

$$\vec{w}_{i_1} = se^{i\phi_z}\{\zeta_{j_1}\vec{p}^{\,b}_{\ j_1} + (1 - \zeta_{j_1})\vec{p}^{\,a}_{\ j_1}\} + \vec{t} \tag{F.11}$$

$$\vec{w}_{i_2} = se^{i\phi_z}\{\zeta_{j_2}\vec{p}^{\,b}_{\ j_2} + (1 - \zeta_{j_2})\vec{p}^{\,a}_{\ j_2}\} + \vec{t} \tag{F.12}$$

$$\vec{w}_{i_3} = se^{i\phi_z}\{\zeta_{j_3}\vec{p}^{\,b}_{\ j_3} + (1 - \zeta_{j_3})\vec{p}^{\,a}_{\ j_3}\} + \vec{t} \tag{F.13}$$

$$\vec{w}_{i_4} = se^{i\phi_z}\{\zeta_{j_4}\vec{p}^{\,b}_{\ j_4} + (1 - \zeta_{j_4})\vec{p}^{\,a}_{\ j_4}\} + \vec{t} \tag{F.14}$$

This results in 8 equations in 8 unknowns (four pose parameters, and four $\zeta$'s), and can be solved to find $\mathcal{T}[\cdot]$.

The TTTT constraint equations are solved as follows. Subtracting Equation (F.12) from (F.11), (F.13) from (F.12), and (F.14) from (F.13) we get

$$\vec{w}_{i_1} - \vec{w}_{i_2} = se^{i\phi_z}\{\zeta_{j_1}(\vec{p}^{\,b}_{\ j_1} - \vec{p}^{\,a}_{\ j_1}) - \zeta_{j_2}(\vec{p}^{\,b}_{\ j_2} - \vec{p}^{\,a}_{\ j_2}) + \vec{p}^{\,a}_{\ j_1} - \vec{p}^{\,a}_{\ j_2}\}$$

$$\vec{w}_{i_2} - \vec{w}_{i_3} = se^{i\phi_z}\{\zeta_{j_2}(\vec{p}^{\,b}_{\ j_2} - \vec{p}^{\,a}_{\ j_2}) - \zeta_{j_3}(\vec{p}^{\,b}_{\ j_3} - \vec{p}^{\,a}_{\ j_3}) + \vec{p}^{\,a}_{\ j_2} - \vec{p}^{\,a}_{\ j_3}\}$$

$$\vec{w}_{i_3} - \vec{w}_{i_4} = se^{i\phi_z}\{\zeta_{j_3}(\vec{p}^{\,b}_{\ j_3} - \vec{p}^{\,a}_{\ j_3}) - \zeta_{j_4}(\vec{p}^{\,b}_{\ j_4} - \vec{p}^{\,a}_{\ j_4}) + \vec{p}^{\,a}_{\ j_3} - \vec{p}^{\,a}_{\ j_4}\}$$

$$\Rightarrow se^{i\phi_z} = \frac{\vec{w}_{i_1} - \vec{w}_{i_2}}{\{\zeta_{j_1}(\vec{p}^{\,b}_{\ j_1} - \vec{p}^{\,a}_{\ j_1}) - \zeta_{j_2}(\vec{p}^{\,b}_{\ j_2} - \vec{p}^{\,a}_{\ j_2}) + \vec{p}^{\,a}_{\ j_1} - \vec{p}^{\,a}_{\ j_2}\}}$$

$$= \frac{\vec{w}_{i_2} - \vec{w}_{i_3}}{\{\zeta_{j_2}(\vec{p}^{\,b}_{\ j_2} - \vec{p}^{\,a}_{\ j_2}) - \zeta_{j_3}(\vec{p}^{\,b}_{\ j_3} - \vec{p}^{\,a}_{\ j_3}) + \vec{p}^{\,a}_{\ j_2} - \vec{p}^{\,a}_{\ j_3}\}}$$

$$= \frac{\vec{w}_{i_3} - \vec{w}_{i_4}}{\{\zeta_{j_3}(\vec{p}^{\,b}_{\ j_3} - \vec{p}^{\,a}_{\ j_3}) - \zeta_{j_4}(\vec{p}^{\,b}_{\ j_4} - \vec{p}^{\,a}_{\ j_4}) + \vec{p}^{\,a}_{\ j_3} - \vec{p}^{\,a}_{\ j_4}\}} \tag{F.15}$$

Equations (F.15) give four real equations in the four real unknowns $\zeta_{j_1}$, $\zeta_{j_2}$, $\zeta_{j_3}$, and $\zeta_{j_4}$. They form a system of *linear equations*, as can be seen by rearranging the first and second, and the second and third terms in Equation (F.15):

$$\text{Re}[(\vec{w}_{i_1} - \vec{w}_{i_2}) \cdot \{\zeta_{j_2}(\vec{p}^{\,b}_{\ j_2} - \vec{p}^{\,a}_{\ j_2}) - \zeta_{j_3}(\vec{p}^{\,b}_{\ j_3} - \vec{p}^{\,a}_{\ j_3}) + \vec{p}^{\,a}_{\ j_2} - \vec{p}^{\,a}_{\ j_3}\}]$$
$$= \text{Re}[(\vec{w}_{i_2} - \vec{w}_{i_3}) \cdot \{\zeta_{j_1}(\vec{p}^{\,b}_{\ j_1} - \vec{p}^{\,a}_{\ j_1}) - \zeta_{j_2}(\vec{p}^{\,b}_{\ j_2} - \vec{p}^{\,a}_{\ j_2}) + \vec{p}^{\,a}_{\ j_1} - \vec{p}^{\,a}_{\ j_2}\}]$$

$$\text{Im}[(\vec{w}_{i_1} - \vec{w}_{i_2}) \cdot \{\zeta_{j_2}(\vec{p}^{\,b}_{\ j_2} - \vec{p}^{\,a}_{\ j_2}) - \zeta_{j_3}(\vec{p}^{\,b}_{\ j_3} - \vec{p}^{\,a}_{\ j_3}) + \vec{p}^{\,a}_{\ j_2} - \vec{p}^{\,a}_{\ j_3}\}]$$
$$= \text{Im}[(\vec{w}_{i_2} - \vec{w}_{i_3}) \cdot \{\zeta_{j_1}(\vec{p}^{\,b}_{\ j_1} - \vec{p}^{\,a}_{\ j_1}) - \zeta_{j_2}(\vec{p}^{\,b}_{\ j_2} - \vec{p}^{\,a}_{\ j_2}) + \vec{p}^{\,a}_{\ j_1} - \vec{p}^{\,a}_{\ j_2}\}]$$

$$\text{Re}[(\vec{w}_{i_2} - \vec{w}_{i_3}) \cdot \{\zeta_{j_3}(\vec{p}^{\,b}_{\ j_3} - \vec{p}^{\,a}_{\ j_3}) - \zeta_{j_4}(\vec{p}^{\,b}_{\ j_4} - \vec{p}^{\,a}_{\ j_4}) + \vec{p}^{\,a}_{\ j_3} - \vec{p}^{\,a}_{\ j_4}\}]$$
$$= \text{Re}[(\vec{w}_{i_3} - \vec{w}_{i_4}) \cdot \{\zeta_{j_2}(\vec{p}^{\,b}_{\ j_2} - \vec{p}^{\,a}_{\ j_2}) - \zeta_{j_3}(\vec{p}^{\,b}_{\ j_3} - \vec{p}^{\,a}_{\ j_3}) + \vec{p}^{\,a}_{\ j_2} - \vec{p}^{\,a}_{\ j_3}\}]$$

$$\text{Im}[(\vec{w}_{i_2} - \vec{w}_{i_3}) \cdot \{\zeta_{j_3}(\vec{p}^{\,b}_{\ j_3} - \vec{p}^{\,a}_{\ j_3}) - \zeta_{j_4}(\vec{p}^{\,b}_{\ j_4} - \vec{p}^{\,a}_{\ j_4}) + \vec{p}^{\,a}_{\ j_3} - \vec{p}^{\,a}_{\ j_4}\}]$$
$$= \text{Im}[(\vec{w}_{i_3} - \vec{w}_{i_4}) \cdot \{\zeta_{j_2}(\vec{p}^{\,b}_{\ j_2} - \vec{p}^{\,a}_{\ j_2}) - \zeta_{j_3}(\vec{p}^{\,b}_{\ j_3} - \vec{p}^{\,a}_{\ j_3}) + \vec{p}^{\,a}_{\ j_2} - \vec{p}^{\,a}_{\ j_3}\}]$$

and can be solved for $\zeta_{j_1}$, $\zeta_{j_2}$, $\zeta_{j_3}$, and $\zeta_{j_4}$ using Gaussian elimination. Once $\zeta_{j_1}$, $\zeta_{j_2}$, $\zeta_{j_3}$, and $\zeta_{j_4}$ are found, they are substituted into Equation (F.15) to compute $se^{i\phi_z}$, giving the scale $s$ and rotation $\phi_z$. Substituting $se^{i\phi_z}$ back into Equation (F.11), we get the translation $\vec{t} = \vec{w}_{i_1} - se^{i\phi_z}\{\zeta_{j_1}\vec{p}^{\,b}_{\ j_1} + (1 - \zeta_{j_1})\vec{p}^{\,a}_{\ j_1}\}$. Note that if $\zeta_{j_1}, \zeta_{j_2}, \zeta_{j_3}, \zeta_{j_4} \notin [0,1]$, the TTTT combination is rejected and another one chosen.

## Appendix G

# Computing Three-Dimensional Pose from Grasp Data

The faces of the shape model are obtained by applying the pose transformation to the shape structure, and are denoted by $\mathbf{f}' = \mathcal{T}[\mathbf{f}]$. Expressing the rotation as a 3x3 orthonormal matrix $R$, these are given by

$$\vec{m}' = \mathcal{T}[\vec{m}] = R\vec{m}$$
$$c' = \mathcal{T}[c] = c + \vec{t} \cdot (R\vec{m}).$$

where the primes (') indicate the transformed values.

The *3-on-3 transform* (Figure 11.5 on page 189) is computed as follows.

(1) Align a model face $\mathbf{f}_{j_1}$ to a data face $\mathbf{s}_{i_1}$. This fixes two rotational degrees of freedom and one translational degree of freedom.

$$R\vec{m}_{j_1} = \vec{n}_{i_1} \qquad\qquad \text{(G.1)}$$
$$c_{j_1} + (R\vec{m}_{j_1}) \cdot \vec{t} = d_{i_1} \qquad\qquad \text{(G.2)}$$

(2) Align direction of the line of intersection of the model faces $\mathbf{f}_{j_1}$ and $\mathbf{f}_{j_2}$ with the direction of the line of intersection of the data faces $\mathbf{s}_{i_1}$ and $\mathbf{s}_{i_2}$. Note that the line of intersection of $\mathbf{f}_{j_1}$ and $\mathbf{f}_{j_2}$ may not correspond to a physical edge; it may be regarded as a virtual model edge, being matched to a virtual data edge due to $\mathbf{s}_{i_1}$ and $\mathbf{s}_{i_2}$. This fixes one more rotational degree of freedom.

$$R(\vec{m}_{j_1} \times \vec{m}_{j_2}) = \vec{n}_{i_1} \times \vec{n}_{i_2} \qquad\qquad \text{(G.3)}$$

(3) Align the distance from the origin of the model face $\mathbf{f}_{j_2}$ with the distance from the origin of the data face $\mathbf{s}_{i_2}$. This fixes one more

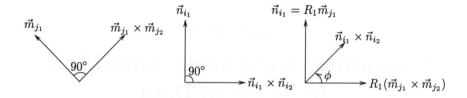

Fig. G.1   Computing the rotation matrix $R$.

translational degree of freedom.

$$c_{j_2} + (R\vec{m}_{j_2}) \cdot \vec{t} = d_{i_2} \qquad (G.4)$$

(4)  Align the distance from the origin of a third model face $\mathbf{f}_{j_3}$ with the distance from the origin of a data face $\mathbf{s}_{i_3}$. This fixes the remaining translational degree of freedom.

$$c_{j_3} + (R\vec{m}_{j_3}) \cdot \vec{t} = d_{i_3} \qquad (G.5)$$

Note that this scheme does not guarantee perfect alignment (exact satisfaction of all three constraint equations) for all the three pairs; only the $\mathbf{f}_{j_1}$ and $\mathbf{s}_{i_1}$ pair is perfectly aligned and gives a zero residual in Equation (11.14). The other two pairs are used to constrain the remaining degrees of freedom. This scheme guarantees at least one residual term to be zero, for a 3-on-3 pose transform. This is in keeping with the observation that the minimal representation cost function tends to favor partial perfect fits.

The expression for a 3-on-3 pose transform is derived by solving the Equations (G.1)–(G.5) above. The rotation matrix $R$ can be obtained by solving Equations (G.1) and (G.3). It can be viewed as a rotation that aligns two pairs of orthogonal unit vectors: the $\vec{n}_{i_1}$ and $\vec{n}_{i_1} \times \vec{n}_{i_2}$ pair in the data, and the $\vec{m}_{j_1}$ and $\vec{m}_{j_1} \times \vec{m}_{j_2}$ pair in the model, as shown in Figure G.1. Thus, $R$ can be found by first aligning the vector $\vec{m}_{j_1}$ with $\vec{n}_{i_1}$; this can be accomplished by rotating $\vec{m}_{j_1}$ about the axis $\vec{m}_{j_1} \times \vec{n}_{i_1}$ by the angle between $\vec{m}_{j_1}$ and $\vec{n}_{i_1}$:

$$R_1 = \text{ROT}(\vec{m}_{j_1} \times \vec{n}_{i_1}, \arccos(\vec{m}_{j_1} \cdot \vec{n}_{i_1})).$$

If $\vec{m}_{j_1}$ with $\vec{n}_{i_1}$ were already aligned, $\vec{m}_{j_1} \times \vec{n}_{i_1} = 0$ and $R_1$ is chosen to be the identity matrix. Once $\vec{m}_{j_1}$ is aligned with $\vec{n}_{i_1}$, rotate the vector $R_1(\vec{m}_{j_1} \times \vec{m}_{j_2})$ about the $\vec{n}_{i_1}$ axis by an angle $\phi$, to align it with the vector

$\vec{n}_{i_1} \times \vec{n}_{i_2}$:

$$R_2 = \text{ROT}(\vec{n}_{i_1}, \phi)$$

where $\phi$ is given by

$$\phi = \arccos((R_1(\vec{m}_{j_1} \times \vec{m}_{j_2})) \cdot (\vec{n}_{i_1} \times \vec{n}_{i_2})),$$

since the rotation axis is now perpendicular to both $R_1(\vec{m}_{j_1} \times \vec{m}_{j_2})$ and $\vec{n}_{i_1} \times \vec{n}_{i_2}$. The rotation matrix $R$ is then given by

$$R = R_2 R_1 = \text{ROT}(\vec{n}_{i_1}, \phi) \cdot \text{ROT}(\vec{m}_{j_1} \times \vec{n}_{i_1}, \arccos(\vec{m}_{j_1} \cdot \vec{n}_{i_1})) \quad \text{(G.6)}$$

Solving Equations (G.2), (G.4), (G.5) for $\vec{t}$ yields

$$\vec{t} = \frac{\left[ \begin{array}{c} (d_{i_1} - c_{j_1})(R\vec{m}_{j_2} \times R\vec{m}_{j_3}) + (d_{i_2} - c_{j_2})(R\vec{m}_{j_3} \times R\vec{m}_{j_1}) + \\ (d_{i_3} - c_{j_3})(R\vec{m}_{j_1} \times R\vec{m}_{j_2}) \end{array} \right]}{R\vec{m}_{j_1} \cdot (R\vec{m}_{j_2} \times R\vec{m}_{j_3})}$$

$$= \frac{\left[ \begin{array}{c} (d_{i_1} - c_{j_1})(R\vec{m}_{j_2} \times R\vec{m}_{j_3}) + (d_{i_2} - c_{j_2})(R\vec{m}_{j_3} \times R\vec{m}_{j_1}) + \\ (d_{i_3} - c_{j_3})(R\vec{m}_{j_1} \times R\vec{m}_{j_2}) \end{array} \right]}{\vec{m}_{j_1} \cdot (\vec{m}_{j_2} \times \vec{m}_{j_3})} \quad \text{(G.7)}$$

where $R$ is the orthonormal rotation matrix found in Equation (G.6). It is possible that a solution to Equations (G.1)–(G.5) does not exist. This happens when the denominator of Equation (G.7) is zero, or when either side of Equation (G.3) is zero. In that case the 3-on-3 pair is rejected and another pair is chosen.

The 3-on-3 pose transform thus found is in a frame obtained by shifting the world coordinate frame to the centroid of the raw contact points (Section 11.2.3.2). This 3-on-3 pose transform is translated back to the original world frame by adding the centroid of the raw contact points. Thus,

$$\vec{t} \leftarrow \vec{t} + <\vec{w}>$$

to get the hypothesized pose in the world coordinates.

# Appendix H

# Quaternion Algebra

## H.1  Definitions

A *quaternion* number is an extension to complex numbers of the form

$$q = [s, \vec{v}] = s + \imath v_x + \jmath v_y + k v_z = r(\cos\phi + \hat{n}\sin\phi) = re^{\hat{n}\phi}$$

where $\imath^2 = \jmath^2 = k^2 = \imath\jmath k = -1$. The *magnitude*, *argument*, and *pole* of $q$ are given by

$$|q| = r = \sqrt{s^2 + \vec{v}\cdot\vec{v}}, \quad \phi = \arctan\frac{|\vec{v}|}{s}, \quad \hat{n} = \frac{\vec{v}}{|\vec{v}|}$$

respectively. A vector $\vec{v}$ is represented by a *pure* quaternion $[0, \vec{v}]$.

The *sum* of two quaternions $q_1$ and $q_2$ is given by

$$q_1 + q_2 = [s_1 + s_2, \vec{v_1} + \vec{v_2}].$$

The *product* of two quaternions $q_1$ and $q_2$ is given by

$$q_1 q_2 = [s_1 s_2 - \vec{v_1}\cdot\vec{v_2}, \; s_1\vec{v_2} + s_2\vec{v_1} + \vec{v_1}\times\vec{v_2}].$$

The *conjugate* and *inverse* of $q$ are defined as

$$q^* = [s, -\vec{v}], \quad \text{and} \quad q^{-1} = \frac{q^*}{|q|^2} = \frac{e^{-\hat{n}\phi}}{r}$$

respectively.

275

## H.2  Properties

All properties of complex numbers are preserved except for commutativity of multiplication. In fact the set of all quaternions, $Q$, together with the sum and product operators comprises a *non-commutative ring with identity and no zero divisors*. Given $q, q_1, q_2, q_3 \in Q$, we have the following.

(1) Sum:

    (a) Closure: $q_1 + q_2 \in Q$
    (b) Commutativity: $q_1 + q_2 = q_2 + q_1$
    (c) Associativity: $(q_1 + q_2) + q_3 = q_1 + (q_2 + q_3)$
    (d) Identity: $\exists 0 \in Q,$ such that $0 + q = q + 0 = q$
    (e) Inverse: $\forall q \in Q, \exists (-q) \in Q,$ such that $q + (-q) = (-q) + q = 0$

(2) Product:

    (a) Closure: $q_1 q_2 \in Q$
    (b) Associativity: $(q_1 q_2) q_3 = q_1 (q_2 q_3)$
    (c) Identity: $\exists 1 \in Q,$ such that $1q = q1 = q$
    (d) Inverse: $\forall q \neq 0 \in Q, \exists q^{-1} \in Q,$ such that $qq^{-1} = q^{-1}q = 1$

(3) Distributivity: $q_1(q_2 + q_3) = q_1 q_2 + q_1 q_3$, and $(q_1 + q_2)q_3 = q_1 q_3 + q_2 q_3$
(4) No zero divisors: $q_1 q_2 = 0 \Rightarrow q_1 = 0$ or $q_2 = 0$

From the above definitions and properties it follows that:

(1) $qq^* = q^*q = |q|^2$
(2) $(q_1 q_2)^* = q_2^* q_1^*$
(3) $|q_1 q_2| = |q_2 q_1| = |q_1| \, |q_2|$
(4) $(q_1 q_2)^{-1} = q_2^{-1} q_1^{-1}$
(5) $q^{\alpha + \beta} = q^\alpha q^\beta$, where $\alpha, \beta \in \Re$
(6) $q^{\alpha \beta} = (q^\alpha)^\beta$, for $|q| \leq 1, \alpha, \beta \in \Re$
(7) $e^{q_1 + q_2} = e^{q_1} e^{q_2} \Leftrightarrow q_1 q_2 = q_2 q_1$
(8) $e^{q_1 q_2} = (e^{q_1})^{q_2} \Leftrightarrow q_1 q_2 = q_2 q_1$

Note that two quaternions, $q_1 = r_1 e^{\hat{n}\phi_1}$ and $q_2 = r_2 e^{\hat{n}\phi_2}$, having the same pole always commute under multiplication

$$r_1 e^{\hat{n}\phi_1} r_2 e^{\hat{n}\phi_2} = r_1 r_2 e^{\hat{n}(\phi_1 + \phi_2)}.$$

Consider pure quaternions (vectors) $\vec{p}_1, \vec{p}_2, \vec{p}_3$. We have $\vec{p}_1 \vec{p}_2 = -\vec{p}_1 \cdot \vec{p}_2 + \vec{p}_1 \times \vec{p}_2$. It is easy to verify the following.

(1) $\vec{p}_1 \cdot \vec{p}_2 = -\frac{\vec{p}_1\vec{p}_2 + \vec{p}_2\vec{p}_1}{2}$

(2) $\vec{p}_1 \times \vec{p}_2 = \frac{\vec{p}_1\vec{p}_2 - \vec{p}_2\vec{p}_1}{2}$

(3) $[\vec{p}_1\ \vec{p}_2\ \vec{p}_3] = \vec{p}_1 \cdot (\vec{p}_2 \times \vec{p}_3) = -\frac{\vec{p}_1\vec{p}_2\vec{p}_3 - \vec{p}_3\vec{p}_2\vec{p}_1}{2}$

(4) $\vec{p}_1 \times (\vec{p}_2 \times \vec{p}_3) = \frac{\vec{p}_1\vec{p}_2\vec{p}_3 - \vec{p}_2\vec{p}_3\vec{p}_1}{2}$

(5) $(\vec{p}_1 \times \vec{p}_2) \times \vec{p}_3 = \frac{\vec{p}_1\vec{p}_2\vec{p}_3 - \vec{p}_3\vec{p}_1\vec{p}_2}{2}$

## H.3   Rotations

The quaternion product $q\vec{p}q^*$, given by

$$q\vec{p}q^* = [0,\ \ 2(\vec{v} \cdot \vec{p})\vec{v} + (s^2 - \vec{v} \cdot \vec{v})\vec{p} + 2s(\vec{v} \times \vec{p})],$$

is the vector $\vec{p}$ rotated around the pole $\hat{n}$ by an angle $2\phi$, and scaled by a factor of $|q|^2$. Therefore, a rotation of $\phi \in [0, 2\pi)$ around an axis $\hat{n}$ is given by a *unit quaternion*,

$$q = [s, \vec{v}] = \cos\frac{\phi}{2} + \hat{n}\sin\frac{\phi}{2} = e^{\hat{n}\frac{\phi}{2}}, \quad |q| = 1.$$

Note that $\phi \pm 2k\pi$ represent the same rotation:

$$e^{\hat{n}\frac{\phi \pm 2k\pi}{2}} = (-1)^k\{\cos(\frac{\phi}{2} + \hat{n}\sin\frac{\phi}{2}\}$$

for $k = 0, 1, \ldots$. Thus, $q$ and $-q$ represent the same rotation, and are periodic in $\phi$, with a period of $4\pi$.

For a unit quaternion the product $q\vec{p}q^*$ simplifies to

$$q\vec{p}q^* = [0,\ \ \vec{p} + 2s(\vec{v} \times \vec{p}) + 2\vec{v} \times (\vec{v} \times \vec{p})], \quad \text{where} \quad |q| = 1,$$

and is equivalent to

$$q\vec{p}q^* = [0,\ \ \vec{p}\cos\phi + (\hat{n} \times \vec{p})\sin\phi + \hat{n}(\hat{n} \cdot \vec{p})(1 - \cos\phi)],$$

which is the same as the well-known *Rodrigues's formula*.

Consider the problem of finding the rotations that map a point $\vec{p}_1$ to the point $\vec{p}_2$. Clearly such a rotation is not unique, and one can isolate the structure of the solution space using unit quaternions. Let the unknown rotation be given by a unit quaternion $q$

$$s^2 + \vec{v} \cdot \vec{v} = 1. \tag{H.1}$$

Then $\vec{p}_2 = q\vec{p}_1 q^*$. Post-multiplying both sides by $q$, gives

$$\vec{p}_2 q = q\vec{p}_1. \tag{H.2}$$

Taking the quaternion products and equating the real and vector parts

$$\vec{v} \cdot (\vec{p}_1 - \vec{p}_2) = 0 \tag{H.3}$$

$$s(\vec{p}_1 - \vec{p}_2) = \vec{v} \times (\vec{p}_1 + \vec{p}_2). \tag{H.4}$$

From Equation (H.3), the axis of rotation or the pole $\hat{n}$, lies in the plane orthogonal to the vector $\vec{p}_1 - \vec{p}_2$.

Given another pair of points, such that $\vec{p}_1'$ maps to $\vec{p}_2'$ under the same rotation, the rotation can be computed as follows. The pole is along the vector $\vec{v} = \gamma(\vec{p}_1 - \vec{p}_2) \times (\vec{p}_1' - \vec{p}_2')$, where $\gamma$ is some constant. Substituting this in Equation (H.4) gives $s$ in terms of $\gamma$. Substituting $s$ and $\vec{v}$ in Equation (H.1) gives $\gamma$, and hence $q$.

# Appendix I

# Computing Three-Dimensional Pose from Vision and Touch Data

## I.1 VVV

The VVV CDFS aligns three arbitrary vision data features $\vec{r}_{i_1}, \vec{r}_{i_2}, \vec{r}_{i_3}$ with three arbitrary model vertex features $\mathcal{T}[\vec{p}_{j_1}], \mathcal{T}[\vec{p}_{j_2}], \mathcal{T}[\vec{p}_{j_3}]$ respectively (Equation 11.6):

$$\vec{e} + \zeta_1 R_c{}^T \vec{r}_{i_1} = q\vec{p}_{j_1} q^* + \vec{t} \tag{I.1}$$

$$\vec{e} + \zeta_2 R_c{}^T \vec{r}_{i_2} = q\vec{p}_{j_2} q^* + \vec{t} \tag{I.2}$$

$$\vec{e} + \zeta_3 R_c{}^T \vec{r}_{i_3} = q\vec{p}_{j_3} q^* + \vec{t} \tag{I.3}$$

This results in 9 real equations in 9 unknowns (six pose parameters, and three $\zeta$'s), and can be solved to find $\mathcal{T}[\cdot]$.

We can eliminate $\vec{t}$ by taking pairwise differences between the three equations. Post-multiplication by $q$ gives:

$$\{\zeta_2 R_c{}^T \vec{r}_{i_2} - \zeta_1 R_c{}^T \vec{r}_{i_1}\} q = q(\vec{p}_{j_2} - \vec{p}_{j_1}) \tag{I.4}$$

$$\{\zeta_3 R_c{}^T \vec{r}_{i_3} - \zeta_2 R_c{}^T \vec{r}_{i_2}\} q = q(\vec{p}_{j_3} - \vec{p}_{j_2}) \tag{I.5}$$

$$\{\zeta_1 R_c{}^T \vec{r}_{i_1} - \zeta_3 R_c{}^T \vec{r}_{i_3}\} q = q(\vec{p}_{j_1} - \vec{p}_{j_3}) \tag{I.6}$$

which is a set of 12 equations in 6 unknowns. We can eliminate $q$ by taking the magnitude of both sides:

$$\{\zeta_2 R_c{}^T \vec{r}_{i_2} - \zeta_1 R_c{}^T \vec{r}_{i_1}\}^2 = (\vec{p}_{j_2} - \vec{p}_{j_1})^2 \tag{I.7}$$

$$\{\zeta_3 R_c{}^T \vec{r}_{i_3} - \zeta_2 R_c{}^T \vec{r}_{i_2}\}^2 = (\vec{p}_{j_3} - \vec{p}_{j_2})^2 \tag{I.8}$$

$$\{\zeta_1 R_c{}^T \vec{r}_{i_1} - \zeta_3 R_c{}^T \vec{r}_{i_3}\}^2 = (\vec{p}_{j_1} - \vec{p}_{j_3})^2 \tag{I.9}$$

where each equation describes an elliptical cylinder in the $\zeta_1, \zeta_2, \zeta_3$ space. The three elliptical cylinders are orthogonal, and in general, intersect at 8 points, giving 8 solutions for $(\zeta_1, \zeta_2, \zeta_3)$. Closed form solutions to the quadratic systems of equations of the form (I.7)-(I.9) are described in the literature (see Linnainmaa et al. [122] and Haralick and Shapiro [77, page 87]). It is also possible to directly solve Equations (I.4)-(I.6) as a the polynomial system of equations, using homotopy based polynomial continuation methods [138].

Of these solutions, only the ones with $\zeta_i \geq 1$ are physically meaningful. Substituting those values back into Equations (I.4) and (I.5), we can find the rotation quaternion $q$, as described in Section H.3[†]. This value of rotation may be substituted into Equation (I.6), to verify that it is valid.

Substituting the value of $q$ and $\zeta_i$ back into Equations (I.1)-(I.3), gives the value of $\vec{t}$. For numerical stability, one can average the values obtained from each of the three equations.

## I.2   VVTT

The VVTT CDFS aligns two arbitrary vision data features $\vec{r}_{i_1}$, $\vec{r}_{i_2}$ and two arbitrary touch data features $\vec{w}_{i_3}$, $\vec{w}_{i_4}$ with two arbitrary model vertex features $\mathcal{T}[\vec{p}_{j_1}]$, $\mathcal{T}[\vec{p}_{j_2}]$ and two arbitrary model face features $\mathcal{T}[\mathbf{f}_{j_3}]$, $\mathcal{T}[\mathbf{f}_{j_4}]$ respectively (Equation 11.6 and Equation 11.9):

$$\vec{e} + \zeta_1 R_c{}^{\mathrm{T}} \vec{r}_{i_1} = q\vec{p}_{j_1} q^* + \vec{t} \tag{I.10}$$

$$\vec{e} + \zeta_2 R_c{}^{\mathrm{T}} \vec{r}_{i_2} = q\vec{p}_{j_2} q^* + \vec{t} \tag{I.11}$$

$$(\vec{w}_{i_3} - \vec{t}) \cdot (q\vec{m}_{j_3} q^*) = c_{j_3} \tag{I.12}$$

$$(\vec{w}_{i_4} - \vec{t}) \cdot (q\vec{m}_{j_4} q^*) = c_{j_4} \tag{I.13}$$

This results in 8 real equations in 8 unknowns (the six pose parameters, and two $\zeta$'s), and can be solved to find $\mathcal{T}[\cdot]$. Alternatively, the last two touch constraints may be written as:

$$\vec{w}_{i_3} = q\{\vec{p}^a{}_{j_3} + \zeta_3(\vec{p}^b{}_{j_3} - \vec{p}^a{}_{j_3}) + \xi_3(\vec{p}^c{}_{j_3} - \vec{p}^a{}_{j_3})\}q^* + \vec{t} \tag{I.14}$$

$$\vec{w}_{i_4} = q\{\vec{p}^a{}_{j_4} + \zeta_4(\vec{p}^b{}_{j_3} - \vec{p}^a{}_{j_4}) + \xi_4(\vec{p}^c{}_{j_3} - \vec{p}^a{}_{j_4})\}q^* + \vec{t} \tag{I.15}$$

[†]Note that Equations (I.4)-(I.6) have the same form as Equation (H.2).

where $\vec{p}^a, \vec{p}^b, \vec{p}^c$ are some three vertices, on the face **f**, and $\zeta, \xi$ are parameters describing points on this face plane. Equations (I.10),(I.11),(I.14),(I.15) give 12 equations in 12 unknowns (six pose parameters, and two pairs of $\zeta, \xi$'s, and two $\zeta$'s).

We can eliminate $\vec{t}$ by taking pairwise differences between the three equations. Post-multiplication by $q$ gives:

$$\{\zeta_2 R_c{}^T \vec{r}_{i_2} - \zeta_1 R_c{}^T \vec{r}_{i_1}\}q = q(\vec{p}_{j_2} - \vec{p}_{j_1}) \tag{I.16}$$

$$\{\vec{w}_{i_3} - \zeta_1 R_c{}^T \vec{r}_{i_1}\}q = q\{\vec{p}^a{}_{j_3} + \zeta_3(\vec{p}^b{}_{j_3} - \vec{p}^a{}_{j_3}) + \xi_3(\vec{p}^c{}_{j_3} - \vec{p}^a{}_{j_3}) - \vec{p}_{j_1}\} \tag{I.17}$$

$$\{\vec{w}_{i_3} - \zeta_2 R_c{}^T \vec{r}_{i_2}\}q = q\{\vec{p}^a{}_{j_3} + \zeta_3(\vec{p}^b{}_{j_3} - \vec{p}^a{}_{j_3}) + \xi_3(\vec{p}^c{}_{j_3} - \vec{p}^a{}_{j_3}) - \vec{p}_{j_2}\} \tag{I.18}$$

$$\{\vec{w}_{i_4} - \zeta_1 R_c{}^T \vec{r}_{i_1}\}q = q\{\vec{p}^a{}_{j_4} + \zeta_4(\vec{p}^b{}_{j_4} - \vec{p}^a{}_{j_4}) + \xi_4(\vec{p}^c{}_{j_4} - \vec{p}^a{}_{j_4}) - \vec{p}_{j_1}\} \tag{I.19}$$

$$\{\vec{w}_{i_4} - \zeta_2 R_c{}^T \vec{r}_{i_2}\}q = q\{\vec{p}^a{}_{j_4} + \zeta_4(\vec{p}^b{}_{j_4} - \vec{p}^a{}_{j_4}) + \xi_4(\vec{p}^c{}_{j_4} - \vec{p}^a{}_{j_4}) - \vec{p}_{j_2}\} \tag{I.20}$$

$$\{\vec{w}_{i_4} - \vec{w}_{i_3}\}q = q\{\vec{p}^a{}_{j_4} + \zeta_4(\vec{p}^b{}_{j_4} - \vec{p}^a{}_{j_4}) + \xi_4(\vec{p}^c{}_{j_4} - \vec{p}^a{}_{j_4}) - \vec{p}^a{}_{j_3} - \zeta_3(\vec{p}^b{}_{j_3} - \vec{p}^a{}_{j_3}) - \xi_3(\vec{p}^c{}_{j_3} - \vec{p}^a{}_{j_3})\} \tag{I.21}$$

which is a set of 24 real equations in 9 unknowns. We can eliminate $q$ by

taking the magnitude of both sides:

$$\{\zeta_2 R_c^{\mathrm{T}} \vec{r}_{i_2} - \zeta_1 R_c^{\mathrm{T}} \vec{r}_{i_1}\}^2 = (\vec{p}_{j_2} - \vec{p}_{j_1})^2 \tag{I.22}$$

$$\{\vec{w}_{i_3} - \zeta_1 R_c^{\mathrm{T}} \vec{r}_{i_1}\}^2 = \{\vec{p}^{\,a}{}_{j_3} + \zeta_3(\vec{p}^{\,b}{}_{j_3} - \vec{p}^{\,a}{}_{j_3}) + \\ \xi_3(\vec{p}^{\,c}{}_{j_3} - \vec{p}^{\,a}{}_{j_3}) - \vec{p}_{j_1}\}^2 \tag{I.23}$$

$$\{\vec{w}_{i_3} - \zeta_2 R_c^{\mathrm{T}} \vec{r}_{i_2}\}^2 = \{\vec{p}^{\,a}{}_{j_3} + \zeta_3(\vec{p}^{\,b}{}_{j_3} - \vec{p}^{\,a}{}_{j_3}) + \\ \xi_3(\vec{p}^{\,c}{}_{j_3} - \vec{p}^{\,a}{}_{j_3}) - \vec{p}_{j_2}\}^2 \tag{I.24}$$

$$\{\vec{w}_{i_4} - \zeta_1 R_c^{\mathrm{T}} \vec{r}_{i_1}\}^2 = \{\vec{p}^{\,a}{}_{j_4} + \zeta_4(\vec{p}^{\,b}{}_{j_4} - \vec{p}^{\,a}{}_{j_4}) + \\ \xi_4(\vec{p}^{\,c}{}_{j_4} - \vec{p}^{\,a}{}_{j_4}) - \vec{p}_{j_1}\}^2 \tag{I.25}$$

$$\{\vec{w}_{i_4} - \zeta_2 R_c^{\mathrm{T}} \vec{r}_{i_2}\}^2 = \{\vec{p}^{\,a}{}_{j_4} + \zeta_4(\vec{p}^{\,b}{}_{j_4} - \vec{p}^{\,a}{}_{j_4}) + \\ \xi_4(\vec{p}^{\,c}{}_{j_4} - \vec{p}^{\,a}{}_{j_4}) - \vec{p}_{j_2}\}^2 \tag{I.26}$$

$$\{\vec{w}_{i_4} - \vec{w}_{i_3}\}^2 = \{\vec{p}^{\,a}{}_{j_4} + \zeta_4(\vec{p}^{\,b}{}_{j_4} - \vec{p}^{\,a}{}_{j_4}) + \xi_4(\vec{p}^{\,c}{}_{j_4} - \vec{p}^{\,a}{}_{j_4}) \\ - \vec{p}^{\,a}{}_{j_3} - \zeta_3(\vec{p}^{\,b}{}_{j_3} - \vec{p}^{\,a}{}_{j_3}) - \xi_3(\vec{p}^{\,c}{}_{j_3} - \vec{p}^{\,a}{}_{j_3})\}^2 \tag{I.27}$$

giving us 6 quadratics in 6 unknowns in the $\zeta_1, \zeta_2, \zeta_3, \xi_3, \zeta_4, \xi_4$ space. In general there are $2^6 = 64$ solutions to such a system of equations. Homotopy based polynomial continuation methods [138] may be used to find all the solutions to such a system of equations.

Of these solutions, only the ones with $\zeta_i \geq 1$, and $\zeta_i, \xi_i$ such that the contact point is contained in the face, are the physically meaningful. Substituting those values back into Equations (I.16) and (I.17), we can find the rotation quaternion $q$, as described in Section H.3[‡]. This value of rotation may be substituted into Equations (I.18)-(I.21), to verify that it is valid.

Substituting the value of $q$ and $\zeta_i, \zeta_i, \xi_i$ back into Equations (I.10), (I.11), (I.14), (I.15) the gives the value of $\vec{t}$. For numerical stability, one can average the values obtained from each of these equations.

Note that the process of elimination can introduce "extra" solutions. There may be other ways to solving this CDFS system of equations, which may result in a smaller number of solutions. Also, further reduction of the quadratic Equations (I.22)-(I.27) may be possible. It is also possible to directly solve Equations (I.16)-(I.21) as a the polynomial system of equations, using homotopy based polynomial continuation methods [138].

---

[‡]Note that Equations (I.16)-(I.21) have the same form as Equation (H.2).

## I.3 VTTTT

The VTTTT CDFS aligns one arbitrary vision data feature $\vec{r}_{i_1}$ and four arbitrary touch data features $\vec{w}_{i_2}, \vec{w}_{i_3}, \vec{w}_{i_4}, \vec{w}_{i_5}$ with one arbitrary model vertex feature $\mathcal{T}[\vec{p}_{j_1}]$ and four arbitrary model face features $\mathcal{T}[\mathbf{f}_{j_2}], \mathcal{T}[\mathbf{f}_{j_3}],$ $\mathcal{T}[\mathbf{f}_{j_4}], \mathcal{T}[\mathbf{f}_{j_5}]$ respectively (Equation 11.6 and Equation 11.9):

$$\vec{e} + \zeta_1 R_c{}^T \vec{r}_{i_1} = q\vec{p}_{j_1}q^* + \vec{t} \tag{I.28}$$

$$(\vec{w}_{i_2} - \vec{t}) \cdot (q\vec{m}_{j_2}q^*) = c_{j_2} \tag{I.29}$$

$$(\vec{w}_{i_3} - \vec{t}) \cdot (q\vec{m}_{j_3}q^*) = c_{j_3} \tag{I.30}$$

$$(\vec{w}_{i_4} - \vec{t}) \cdot (q\vec{m}_{j_4}q^*) = c_{j_4} \tag{I.31}$$

$$(\vec{w}_{i_5} - \vec{t}) \cdot (q\vec{m}_{j_5}q^*) = c_{j_5} \tag{I.32}$$

This results in 7 real equations in 7 unknowns (the six pose parameters, and one $\zeta$), and can be solved to find $\mathcal{T}[\cdot]$. Alternatively, the touch constraints may be written as:

$$\vec{w}_{i_2} = q\{\vec{p}^a{}_{j_2} + \zeta_2(\vec{p}^b{}_{j_2} - \vec{p}^a{}_{j_2}) + \xi_2(\vec{p}^c{}_{j_2} - \vec{p}^a{}_{j_2})\}q^* + \vec{t} \tag{I.33}$$

$$\vec{w}_{i_3} = q\{\vec{p}^a{}_{j_3} + \zeta_3(\vec{p}^b{}_{j_3} - \vec{p}^a{}_{j_3}) + \xi_3(\vec{p}^c{}_{j_3} - \vec{p}^a{}_{j_3})\}q^* + \vec{t} \tag{I.34}$$

$$\vec{w}_{i_4} = q\{\vec{p}^a{}_{j_4} + \zeta_4(\vec{p}^b{}_{j_3} - \vec{p}^a{}_{j_4}) + \xi_4(\vec{p}^c{}_{j_3} - \vec{p}^a{}_{j_4})\}q^* + \vec{t} \tag{I.35}$$

$$\vec{w}_{i_5} = q\{\vec{p}^a{}_{j_5} + \zeta_5(\vec{p}^b{}_{j_5} - \vec{p}^a{}_{j_5}) + \xi_5(\vec{p}^c{}_{j_5} - \vec{p}^a{}_{j_5})\}q^* + \vec{t} \tag{I.36}$$

where $\vec{p}^a, \vec{p}^b, \vec{p}^c$ are some three vertices, on the face $\mathbf{f}$, and $\zeta, \xi$ are parameters describing points on this face plane. Equations (I.28),(I.34)-(I.36) give 15 equations in 15 unknowns (six pose parameters, and four pairs of $\zeta, \xi$'s, and one $\zeta$).

We can eliminate $\vec{t}$ by taking pairwise differences between the five equations. Post-multiplication by $q$ gives:

$$\{\vec{w}_{i_2} - \zeta_1 R_c{}^T \vec{r}_{i_1}\}q = q\{\vec{p}^a{}_{j_2} + \zeta_2(\vec{p}^b{}_{j_2} - \vec{p}^a{}_{j_2}) + \xi_2(\vec{p}^c{}_{j_2} - \vec{p}^a{}_{j_2}) - \vec{p}_{j_1}\}$$
(I.37)

$$\{\vec{w}_{i_3} - \zeta_1 R_c{}^T \vec{r}_{i_1}\}q = q\{\vec{p}^a{}_{j_3} + \zeta_3(\vec{p}^b{}_{j_3} - \vec{p}^a{}_{j_3}) + \xi_3(\vec{p}^c{}_{j_3} - \vec{p}^a{}_{j_3}) - \vec{p}_{j_1}\}$$
(I.38)

$$\{\vec{w}_{i_4} - \zeta_1 R_c{}^T \vec{r}_{i_1}\}q = q\{\vec{p}^a{}_{j_4} + \zeta_4(\vec{p}^b{}_{j_4} - \vec{p}^a{}_{j_4}) + \xi_4(\vec{p}^c{}_{j_4} - \vec{p}^a{}_{j_4}) - \vec{p}_{j_1}\}$$
(I.39)

$$\{\vec{w}_{i_5} - \zeta_1 R_c{}^T \vec{r}_{i_1}\}q = q\{\vec{p}^a{}_{j_5} + \zeta_5(\vec{p}^b{}_{j_5} - \vec{p}^a{}_{j_5}) + \xi_5(\vec{p}^c{}_{j_5} - \vec{p}^a{}_{j_5}) - \vec{p}_{j_1}\}$$
(I.40)

$$\{\vec{w}_{i_3} - \vec{w}_{i_2}\}q = q\{\vec{p}^a{}_{j_3} + \zeta_3(\vec{p}^b{}_{j_3} - \vec{p}^a{}_{j_3}) + \xi_3(\vec{p}^c{}_{j_3} - \vec{p}^a{}_{j_3})$$
$$- \vec{p}^a{}_{j_2} - \zeta_2(\vec{p}^b{}_{j_2} - \vec{p}^a{}_{j_2}) - \xi_2(\vec{p}^c{}_{j_2} - \vec{p}^a{}_{j_2})\}$$
(I.41)

$$\{\vec{w}_{i_4} - \vec{w}_{i_2}\}q = q\{\vec{p}^a{}_{j_4} + \zeta_4(\vec{p}^b{}_{j_4} - \vec{p}^a{}_{j_4}) + \xi_4(\vec{p}^c{}_{j_4} - \vec{p}^a{}_{j_4})$$
$$- \vec{p}^a{}_{j_2} - \zeta_2(\vec{p}^b{}_{j_2} - \vec{p}^a{}_{j_2}) - \xi_2(\vec{p}^c{}_{j_2} - \vec{p}^a{}_{j_2})\}$$
(I.42)

$$\{\vec{w}_{i_5} - \vec{w}_{i_2}\}q = q\{\vec{p}^a{}_{j_5} + \zeta_5(\vec{p}^b{}_{j_5} - \vec{p}^a{}_{j_5}) + \xi_5(\vec{p}^c{}_{j_5} - \vec{p}^a{}_{j_5})$$
$$- \vec{p}^a{}_{j_2} - \zeta_2(\vec{p}^b{}_{j_2} - \vec{p}^a{}_{j_2}) - \xi_2(\vec{p}^c{}_{j_2} - \vec{p}^a{}_{j_2})\}$$
(I.43)

$$\{\vec{w}_{i_4} - \vec{w}_{i_3}\}q = q\{\vec{p}^a{}_{j_4} + \zeta_4(\vec{p}^b{}_{j_4} - \vec{p}^a{}_{j_4}) + \xi_4(\vec{p}^c{}_{j_4} - \vec{p}^a{}_{j_4})$$
$$- \vec{p}^a{}_{j_3} - \zeta_3(\vec{p}^b{}_{j_3} - \vec{p}^a{}_{j_3}) - \xi_3(\vec{p}^c{}_{j_3} - \vec{p}^a{}_{j_3})\}$$
(I.44)

$$\{\vec{w}_{i_5} - \vec{w}_{i_3}\}q = q\{\vec{p}^a{}_{j_5} + \zeta_5(\vec{p}^b{}_{j_5} - \vec{p}^a{}_{j_5}) + \xi_5(\vec{p}^c{}_{j_5} - \vec{p}^a{}_{j_5})$$
$$- \vec{p}^a{}_{j_3} - \zeta_3(\vec{p}^b{}_{j_3} - \vec{p}^a{}_{j_3}) - \xi_3(\vec{p}^c{}_{j_3} - \vec{p}^a{}_{j_3})\}$$
(I.45)

which is a set of 36 real equations in 12 unknowns. We can eliminate $q$ by

taking the magnitude of both sides:

$$\{\vec{w}_{i_2} - \zeta_1 R_c{}^T \vec{r}_{i_1}\}^2 = \{\vec{p}{}^a{}_{j_2} + \zeta_2(\vec{p}{}^b{}_{j_2} - \vec{p}{}^a{}_{j_2}) + \xi_2(\vec{p}{}^c{}_{j_2} - \vec{p}{}^a{}_{j_2}) - \vec{p}{}_{j_1}\}^2$$

(I.46)

$$\{\vec{w}_{i_3} - \zeta_1 R_c{}^T \vec{r}_{i_1}\}^2 = \{\vec{p}{}^a{}_{j_3} + \zeta_3(\vec{p}{}^b{}_{j_3} - \vec{p}{}^a{}_{j_3}) + \xi_3(\vec{p}{}^c{}_{j_3} - \vec{p}{}^a{}_{j_3}) - \vec{p}{}_{j_1}\}^2$$

(I.47)

$$\{\vec{w}_{i_4} - \zeta_1 R_c{}^T \vec{r}_{i_1}\}^2 = \{\vec{p}{}^a{}_{j_4} + \zeta_4(\vec{p}{}^b{}_{j_4} - \vec{p}{}^a{}_{j_4}) + \xi_4(\vec{p}{}^c{}_{j_4} - \vec{p}{}^a{}_{j_4}) - \vec{p}{}_{j_1}\}^2$$

(I.48)

$$\{\vec{w}_{i_5} - \zeta_1 R_c{}^T \vec{r}_{i_1}\}^2 = \{\vec{p}{}^a{}_{j_5} + \zeta_5(\vec{p}{}^b{}_{j_5} - \vec{p}{}^a{}_{j_5}) + \xi_5(\vec{p}{}^c{}_{j_5} - \vec{p}{}^a{}_{j_5}) - \vec{p}{}_{j_1}\}^2$$

(I.49)

$$\{\vec{w}_{i_3} - \vec{w}_{i_2}\}^2 = \{\vec{p}{}^a{}_{j_3} + \zeta_3(\vec{p}{}^b{}_{j_3} - \vec{p}{}^a{}_{j_3}) + \xi_3(\vec{p}{}^c{}_{j_3} - \vec{p}{}^a{}_{j_3})$$
$$- \vec{p}{}^a{}_{j_2} - \zeta_2(\vec{p}{}^b{}_{j_2} - \vec{p}{}^a{}_{j_2}) - \xi_2(\vec{p}{}^c{}_{j_2} - \vec{p}{}^a{}_{j_2})\}^2$$

(I.50)

$$\{\vec{w}_{i_4} - \vec{w}_{i_2}\}^2 = \{\vec{p}{}^a{}_{j_4} + \zeta_4(\vec{p}{}^b{}_{j_4} - \vec{p}{}^a{}_{j_4}) + \xi_4(\vec{p}{}^c{}_{j_4} - \vec{p}{}^a{}_{j_4})$$
$$- \vec{p}{}^a{}_{j_2} - \zeta_2(\vec{p}{}^b{}_{j_2} - \vec{p}{}^a{}_{j_2}) - \xi_2(\vec{p}{}^c{}_{j_2} - \vec{p}{}^a{}_{j_2})\}^2$$

(I.51)

$$\{\vec{w}_{i_5} - \vec{w}_{i_2}\}^2 = \{\vec{p}{}^a{}_{j_5} + \zeta_5(\vec{p}{}^b{}_{j_5} - \vec{p}{}^a{}_{j_5}) + \xi_5(\vec{p}{}^c{}_{j_5} - \vec{p}{}^a{}_{j_5})$$
$$- \vec{p}{}^a{}_{j_2} - \zeta_2(\vec{p}{}^b{}_{j_2} - \vec{p}{}^a{}_{j_2}) - \xi_2(\vec{p}{}^c{}_{j_2} - \vec{p}{}^a{}_{j_2})\}^2$$

(I.52)

$$\{\vec{w}_{i_4} - \vec{w}_{i_3}\}^2 = \{\vec{p}{}^a{}_{j_4} + \zeta_4(\vec{p}{}^b{}_{j_4} - \vec{p}{}^a{}_{j_4}) + \xi_4(\vec{p}{}^c{}_{j_4} - \vec{p}{}^a{}_{j_4})$$
$$- \vec{p}{}^a{}_{j_3} - \zeta_3(\vec{p}{}^b{}_{j_3} - \vec{p}{}^a{}_{j_3}) - \xi_3(\vec{p}{}^c{}_{j_3} - \vec{p}{}^a{}_{j_3})\}^2$$

(I.53)

$$\{\vec{w}_{i_5} - \vec{w}_{i_3}\}^2 = \{\vec{p}{}^a{}_{j_5} + \zeta_5(\vec{p}{}^b{}_{j_5} - \vec{p}{}^a{}_{j_5}) + \xi_5(\vec{p}{}^c{}_{j_5} - \vec{p}{}^a{}_{j_5})$$
$$- \vec{p}{}^a{}_{j_3} - \zeta_3(\vec{p}{}^b{}_{j_3} - \vec{p}{}^a{}_{j_3}) - \xi_3(\vec{p}{}^c{}_{j_3} - \vec{p}{}^a{}_{j_3})\}^2$$

(I.54)

giving us 9 quadratics in 9 unknowns in the $\zeta_1, \zeta_2, \xi_2, \zeta_3, \xi_3, \zeta_4, \xi_4, \zeta_5, \xi_5$ space. In general there are $2^9 = 512$ solutions to such a system of equations. Homotopy based polynomial continuation methods [138] may be used to find all the solutions to such a system of equations.

Of these solutions, only the ones with $\zeta_i \geq 1$, and $\zeta_i, \xi_i$ such that the contact point is contained in the face, are the physically meaningful. Substituting those values back into Equations (I.37) and (I.38), we can find the rotation quaternion $q$, as described in Section H.3[§]. This value of rotation may be substituted into Equations (I.39)-(I.45), to verify that it is valid.

Substituting the value of $q$ and $\zeta_i, \zeta_i, \xi_i$ back into Equations (I.28),(I.33)-(I.36) the gives the value of $\vec{t}$. For numerical stability, one can average the

[§]Note that Equations (I.37)-(I.45) have the same form as Equation (H.2).

values obtained from each of these equations.

  Note that the process of elimination can introduce "extra" solutions, especially since we have 9 unknowns and 36 equations. There may be other ways to solving this CDFS system of equations, which may result in a smaller number of solutions. Also, further reduction of the quadratic Equations (I.46)-(I.51) may be possible. It is also possible to directly solve Equations (I.37)-(I.45) as a the polynomial system of equations, using homotopy based polynomial continuation methods [138].

## I.4  TTTTTT

The TTTTTT CDFS aligns six arbitrary touch data features $\vec{w}_{i_1}$, $\vec{w}_{i_2}$, $\vec{w}_{i_3}$, $\vec{w}_{i_4}$, $\vec{w}_{i_5}$, $\vec{w}_{i_6}$ with six arbitrary model face features $\mathcal{T}[\mathbf{f}_{j_1}]$, $\mathcal{T}[\mathbf{f}_{j_2}]$, $\mathcal{T}[\mathbf{f}_{j_3}]$, $\mathcal{T}[\mathbf{f}_{j_4}]$, $\mathcal{T}[\mathbf{f}_{j_5}]$, $\mathcal{T}[\mathbf{f}_{j_6}]$ respectively (Equation 11.9):

$$(\vec{w}_{i_1} - \vec{t}) \cdot (q\vec{m}_{j_1}q^*) = c_{j_1} \tag{I.55}$$

$$(\vec{w}_{i_2} - \vec{t}) \cdot (q\vec{m}_{j_2}q^*) = c_{j_2} \tag{I.56}$$

$$(\vec{w}_{i_3} - \vec{t}) \cdot (q\vec{m}_{j_3}q^*) = c_{j_3} \tag{I.57}$$

$$(\vec{w}_{i_4} - \vec{t}) \cdot (q\vec{m}_{j_4}q^*) = c_{j_4} \tag{I.58}$$

$$(\vec{w}_{i_5} - \vec{t}) \cdot (q\vec{m}_{j_5}q^*) = c_{j_5} \tag{I.59}$$

$$(\vec{w}_{i_6} - \vec{t}) \cdot (q\vec{m}_{j_6}q^*) = c_{j_6} \tag{I.60}$$

This results in 6 real equations in 6 unknowns (the six pose parameters), and can be solved to find $\mathcal{T}[\cdot]$. Alternatively, the above touch constraints may be written as:

$$\vec{w}_{i_1} = q\{\vec{p}^a_{\ j_1} + \zeta_1(\vec{p}^b_{\ j_1} - \vec{p}^a_{\ j_1}) + \xi_1(\vec{p}^c_{\ j_1} - \vec{p}^a_{\ j_1})\}q^* + \vec{t} \tag{I.61}$$

$$\vec{w}_{i_2} = q\{\vec{p}^a_{\ j_2} + \zeta_2(\vec{p}^b_{\ j_2} - \vec{p}^a_{\ j_2}) + \xi_2(\vec{p}^c_{\ j_2} - \vec{p}^a_{\ j_2})\}q^* + \vec{t} \tag{I.62}$$

$$\vec{w}_{i_3} = q\{\vec{p}^a_{\ j_3} + \zeta_3(\vec{p}^b_{\ j_3} - \vec{p}^a_{\ j_3}) + \xi_3(\vec{p}^c_{\ j_3} - \vec{p}^a_{\ j_3})\}q^* + \vec{t} \tag{I.63}$$

$$\vec{w}_{i_4} = q\{\vec{p}^a_{\ j_4} + \zeta_4(\vec{p}^b_{\ j_3} - \vec{p}^a_{\ j_4}) + \xi_4(\vec{p}^c_{\ j_3} - \vec{p}^a_{\ j_4})\}q^* + \vec{t} \tag{I.64}$$

$$\vec{w}_{i_5} = q\{\vec{p}^a_{\ j_5} + \zeta_5(\vec{p}^b_{\ j_5} - \vec{p}^a_{\ j_5}) + \xi_5(\vec{p}^c_{\ j_5} - \vec{p}^a_{\ j_5})\}q^* + \vec{t} \tag{I.65}$$

$$\vec{w}_{i_6} = q\{\vec{p}^a_{\ j_6} + \zeta_6(\vec{p}^b_{\ j_6} - \vec{p}^a_{\ j_6}) + \xi_6(\vec{p}^c_{\ j_6} - \vec{p}^a_{\ j_6})\}q^* + \vec{t} \tag{I.66}$$

where $\vec{p}^a, \vec{p}^b, \vec{p}^c$ are some three vertices, on the face $\mathbf{f}$, and $\zeta, \xi$ are parameters describing points on this face plane. Equations (I.61)-(I.66) give 18 equations in 18 unknowns (six pose parameters, and six pairs of $\zeta, \xi$'s).

We can eliminate $\vec{t}$ by taking pairwise differences between the six equations. Post-multiplication by $q$ gives:

$$\{\vec{w}_{i_2} - \vec{w}_{i_1}\}q = q\{\vec{p}^a{}_{j_2} + \zeta_2(\vec{p}^b{}_{j_2} - \vec{p}^a{}_{j_2}) + \xi_2(\vec{p}^c{}_{j_2} - \vec{p}^a{}_{j_2})$$
$$- \vec{p}^a{}_{j_1} - \zeta_1(\vec{p}^b{}_{j_1} - \vec{p}^a{}_{j_1}) - \xi_1(\vec{p}^c{}_{j_1} - \vec{p}^a{}_{j_1})\} \qquad (I.67)$$

$$\{\vec{w}_{i_3} - \vec{w}_{i_1}\}q = q\{\vec{p}^a{}_{j_3} + \zeta_3(\vec{p}^b{}_{j_3} - \vec{p}^a{}_{j_3}) + \xi_3(\vec{p}^c{}_{j_3} - \vec{p}^a{}_{j_3})$$
$$- \vec{p}^a{}_{j_1} - \zeta_1(\vec{p}^b{}_{j_1} - \vec{p}^a{}_{j_1}) - \xi_1(\vec{p}^c{}_{j_1} - \vec{p}^a{}_{j_1})\} \qquad (I.68)$$

$$\{\vec{w}_{i_4} - \vec{w}_{i_1}\}q = q\{\vec{p}^a{}_{j_4} + \zeta_4(\vec{p}^b{}_{j_4} - \vec{p}^a{}_{j_4}) + \xi_4(\vec{p}^c{}_{j_4} - \vec{p}^a{}_{j_4})$$
$$- \vec{p}^a{}_{j_1} - \zeta_1(\vec{p}^b{}_{j_1} - \vec{p}^a{}_{j_1}) - \xi_1(\vec{p}^c{}_{j_1} - \vec{p}^a{}_{j_1})\} \qquad (I.69)$$

$$\{\vec{w}_{i_5} - \vec{w}_{i_1}\}q = q\{\vec{p}^a{}_{j_5} + \zeta_5(\vec{p}^b{}_{j_5} - \vec{p}^a{}_{j_5}) + \xi_5(\vec{p}^c{}_{j_5} - \vec{p}^a{}_{j_5})$$
$$- \vec{p}^a{}_{j_1} - \zeta_1(\vec{p}^b{}_{j_1} - \vec{p}^a{}_{j_1}) - \xi_1(\vec{p}^c{}_{j_1} - \vec{p}^a{}_{j_1})\} \qquad (I.70)$$

$$\{\vec{w}_{i_6} - \vec{w}_{i_1}\}q = q\{\vec{p}^a{}_{j_6} + \zeta_6(\vec{p}^b{}_{j_6} - \vec{p}^a{}_{j_6}) + \xi_6(\vec{p}^c{}_{j_6} - \vec{p}^a{}_{j_6})$$
$$- \vec{p}^a{}_{j_1} - \zeta_1(\vec{p}^b{}_{j_1} - \vec{p}^a{}_{j_1}) - \xi_1(\vec{p}^c{}_{j_1} - \vec{p}^a{}_{j_1})\} \qquad (I.71)$$

$$\{\vec{w}_{i_3} - \vec{w}_{i_2}\}q = q\{\vec{p}^a{}_{j_3} + \zeta_3(\vec{p}^b{}_{j_3} - \vec{p}^a{}_{j_3}) + \xi_3(\vec{p}^c{}_{j_3} - \vec{p}^a{}_{j_3})$$
$$- \vec{p}^a{}_{j_2} - \zeta_2(\vec{p}^b{}_{j_2} - \vec{p}^a{}_{j_2}) - \xi_2(\vec{p}^c{}_{j_2} - \vec{p}^a{}_{j_2})\} \qquad (I.72)$$

$$\{\vec{w}_{i_4} - \vec{w}_{i_2}\}q = q\{\vec{p}^a{}_{j_4} + \zeta_4(\vec{p}^b{}_{j_4} - \vec{p}^a{}_{j_4}) + \xi_4(\vec{p}^c{}_{j_4} - \vec{p}^a{}_{j_4})$$
$$- \vec{p}^a{}_{j_2} - \zeta_2(\vec{p}^b{}_{j_2} - \vec{p}^a{}_{j_2}) - \xi_2(\vec{p}^c{}_{j_2} - \vec{p}^a{}_{j_2})\} \qquad (I.73)$$

$$\{\vec{w}_{i_5} - \vec{w}_{i_2}\}q = q\{\vec{p}^a{}_{j_5} + \zeta_5(\vec{p}^b{}_{j_5} - \vec{p}^a{}_{j_5}) + \xi_5(\vec{p}^c{}_{j_5} - \vec{p}^a{}_{j_5})$$
$$- \vec{p}^a{}_{j_2} - \zeta_2(\vec{p}^b{}_{j_2} - \vec{p}^a{}_{j_2}) - \xi_2(\vec{p}^c{}_{j_2} - \vec{p}^a{}_{j_2})\} \qquad (I.74)$$

$$\{\vec{w}_{i_6} - \vec{w}_{i_2}\}q = q\{\vec{p}^a{}_{j_6} + \zeta_6(\vec{p}^b{}_{j_6} - \vec{p}^a{}_{j_6}) + \xi_6(\vec{p}^c{}_{j_6} - \vec{p}^a{}_{j_6})$$
$$- \vec{p}^a{}_{j_2} - \zeta_2(\vec{p}^b{}_{j_2} - \vec{p}^a{}_{j_2}) - \xi_2(\vec{p}^c{}_{j_2} - \vec{p}^a{}_{j_2})\} \qquad (I.75)$$

$$\{\vec{w}_{i_4} - \vec{w}_{i_3}\}q = q\{\vec{p}^a{}_{j_4} + \zeta_4(\vec{p}^b{}_{j_4} - \vec{p}^a{}_{j_4}) + \xi_4(\vec{p}^c{}_{j_4} - \vec{p}^a{}_{j_4})$$
$$- \vec{p}^a{}_{j_3} - \zeta_3(\vec{p}^b{}_{j_3} - \vec{p}^a{}_{j_3}) - \xi_3(\vec{p}^c{}_{j_3} - \vec{p}^a{}_{j_3})\} \qquad (I.76)$$

$$\{\vec{w}_{i_5} - \vec{w}_{i_3}\}q = q\{\vec{p}^a{}_{j_5} + \zeta_5(\vec{p}^b{}_{j_5} - \vec{p}^a{}_{j_5}) + \xi_5(\vec{p}^c{}_{j_5} - \vec{p}^a{}_{j_5})$$
$$- \vec{p}^a{}_{j_3} - \zeta_3(\vec{p}^b{}_{j_3} - \vec{p}^a{}_{j_3}) - \xi_3(\vec{p}^c{}_{j_3} - \vec{p}^a{}_{j_3})\} \qquad (I.77)$$

$$\{\vec{w}_{i_6} - \vec{w}_{i_3}\}q = q\{\vec{p}^a{}_{j_6} + \zeta_6(\vec{p}^b{}_{j_6} - \vec{p}^a{}_{j_6}) + \xi_6(\vec{p}^c{}_{j_6} - \vec{p}^a{}_{j_6})$$
$$- \vec{p}^a{}_{j_3} - \zeta_3(\vec{p}^b{}_{j_3} - \vec{p}^a{}_{j_3}) - \xi_3(\vec{p}^c{}_{j_3} - \vec{p}^a{}_{j_3})\} \qquad (I.78)$$

which is a set of 48 real equations in 15 unknowns. We can eliminate $q$ by

taking the magnitude of both sides:

$$\{\vec{w}_{i_2} - \vec{w}_{i_1}\}^2 = \{\vec{p}^a{}_{j_2} + \zeta_2(\vec{p}^b{}_{j_2} - \vec{p}^a{}_{j_2}) + \xi_2(\vec{p}^c{}_{j_2} - \vec{p}^a{}_{j_2})$$
$$- \vec{p}^a{}_{j_1} - \zeta_1(\vec{p}^b{}_{j_1} - \vec{p}^a{}_{j_1}) - \xi_1(\vec{p}^c{}_{j_1} - \vec{p}^a{}_{j_1})\}^2 \qquad (I.79)$$

$$\{\vec{w}_{i_3} - \vec{w}_{i_1}\}^2 = \{\vec{p}^a{}_{j_3} + \zeta_3(\vec{p}^b{}_{j_3} - \vec{p}^a{}_{j_3}) + \xi_3(\vec{p}^c{}_{j_3} - \vec{p}^a{}_{j_3})$$
$$- \vec{p}^a{}_{j_1} - \zeta_1(\vec{p}^b{}_{j_1} - \vec{p}^a{}_{j_1}) - \xi_1(\vec{p}^c{}_{j_1} - \vec{p}^a{}_{j_1})\}^2 \qquad (I.80)$$

$$\{\vec{w}_{i_4} - \vec{w}_{i_1}\}^2 = \{\vec{p}^a{}_{j_4} + \zeta_4(\vec{p}^b{}_{j_4} - \vec{p}^a{}_{j_4}) + \xi_4(\vec{p}^c{}_{j_4} - \vec{p}^a{}_{j_4})$$
$$- \vec{p}^a{}_{j_1} - \zeta_1(\vec{p}^b{}_{j_1} - \vec{p}^a{}_{j_1}) - \xi_1(\vec{p}^c{}_{j_1} - \vec{p}^a{}_{j_1})\}^2 \qquad (I.81)$$

$$\{\vec{w}_{i_5} - \vec{w}_{i_1}\}^2 = \{\vec{p}^a{}_{j_5} + \zeta_5(\vec{p}^b{}_{j_5} - \vec{p}^a{}_{j_5}) + \xi_5(\vec{p}^c{}_{j_5} - \vec{p}^a{}_{j_5})$$
$$- \vec{p}^a{}_{j_1} - \zeta_1(\vec{p}^b{}_{j_1} - \vec{p}^a{}_{j_1}) - \xi_1(\vec{p}^c{}_{j_1} - \vec{p}^a{}_{j_1})\}^2 \qquad (I.82)$$

$$\{\vec{w}_{i_6} - \vec{w}_{i_1}\}^2 = \{\vec{p}^a{}_{j_6} + \zeta_6(\vec{p}^b{}_{j_6} - \vec{p}^a{}_{j_6}) + \xi_6(\vec{p}^c{}_{j_6} - \vec{p}^a{}_{j_6})$$
$$- \vec{p}^a{}_{j_1} - \zeta_1(\vec{p}^b{}_{j_1} - \vec{p}^a{}_{j_1}) - \xi_1(\vec{p}^c{}_{j_1} - \vec{p}^a{}_{j_1})\}^2 \qquad (I.83)$$

$$\{\vec{w}_{i_3} - \vec{w}_{i_2}\}^2 = \{\vec{p}^a{}_{j_3} + \zeta_3(\vec{p}^b{}_{j_3} - \vec{p}^a{}_{j_3}) + \xi_3(\vec{p}^c{}_{j_3} - \vec{p}^a{}_{j_3})$$
$$- \vec{p}^a{}_{j_2} - \zeta_2(\vec{p}^b{}_{j_2} - \vec{p}^a{}_{j_2}) - \xi_2(\vec{p}^c{}_{j_2} - \vec{p}^a{}_{j_2})\}^2 \qquad (I.84)$$

$$\{\vec{w}_{i_4} - \vec{w}_{i_2}\}^2 = \{\vec{p}^a{}_{j_4} + \zeta_4(\vec{p}^b{}_{j_4} - \vec{p}^a{}_{j_4}) + \xi_4(\vec{p}^c{}_{j_4} - \vec{p}^a{}_{j_4})$$
$$- \vec{p}^a{}_{j_2} - \zeta_2(\vec{p}^b{}_{j_2} - \vec{p}^a{}_{j_2}) - \xi_2(\vec{p}^c{}_{j_2} - \vec{p}^a{}_{j_2})\}^2 \qquad (I.85)$$

$$\{\vec{w}_{i_5} - \vec{w}_{i_2}\}^2 = \{\vec{p}^a{}_{j_5} + \zeta_5(\vec{p}^b{}_{j_5} - \vec{p}^a{}_{j_5}) + \xi_5(\vec{p}^c{}_{j_5} - \vec{p}^a{}_{j_5})$$
$$- \vec{p}^a{}_{j_2} - \zeta_2(\vec{p}^b{}_{j_2} - \vec{p}^a{}_{j_2}) - \xi_2(\vec{p}^c{}_{j_2} - \vec{p}^a{}_{j_2})\}^2 \qquad (I.86)$$

$$\{\vec{w}_{i_6} - \vec{w}_{i_2}\}^2 = \{\vec{p}^a{}_{j_6} + \zeta_6(\vec{p}^b{}_{j_6} - \vec{p}^a{}_{j_6}) + \xi_6(\vec{p}^c{}_{j_6} - \vec{p}^a{}_{j_6})$$
$$- \vec{p}^a{}_{j_2} - \zeta_2(\vec{p}^b{}_{j_2} - \vec{p}^a{}_{j_2}) - \xi_2(\vec{p}^c{}_{j_2} - \vec{p}^a{}_{j_2})\}^2 \qquad (I.87)$$

$$\{\vec{w}_{i_4} - \vec{w}_{i_3}\}^2 = \{\vec{p}^a{}_{j_4} + \zeta_4(\vec{p}^b{}_{j_4} - \vec{p}^a{}_{j_4}) + \xi_4(\vec{p}^c{}_{j_4} - \vec{p}^a{}_{j_4})$$
$$- \vec{p}^a{}_{j_3} - \zeta_3(\vec{p}^b{}_{j_3} - \vec{p}^a{}_{j_3}) - \xi_3(\vec{p}^c{}_{j_3} - \vec{p}^a{}_{j_3})\}^2 \qquad (I.88)$$

$$\{\vec{w}_{i_5} - \vec{w}_{i_3}\}^2 = \{\vec{p}^a{}_{j_5} + \zeta_5(\vec{p}^b{}_{j_5} - \vec{p}^a{}_{j_5}) + \xi_5(\vec{p}^c{}_{j_5} - \vec{p}^a{}_{j_5})$$
$$- \vec{p}^a{}_{j_3} - \zeta_3(\vec{p}^b{}_{j_3} - \vec{p}^a{}_{j_3}) - \xi_3(\vec{p}^c{}_{j_3} - \vec{p}^a{}_{j_3})\}^2 \qquad (I.89)$$

$$\{\vec{w}_{i_6} - \vec{w}_{i_3}\}^2 = \{\vec{p}^a{}_{j_6} + \zeta_6(\vec{p}^b{}_{j_6} - \vec{p}^a{}_{j_6}) + \xi_6(\vec{p}^c{}_{j_6} - \vec{p}^a{}_{j_6})$$
$$- \vec{p}^a{}_{j_3} - \zeta_3(\vec{p}^b{}_{j_3} - \vec{p}^a{}_{j_3}) - \xi_3(\vec{p}^c{}_{j_3} - \vec{p}^a{}_{j_3})\}^2 \qquad (I.90)$$

giving us 12 quadratics in 12 unknowns in the $\zeta_1, \xi_1, \zeta_2, \xi_2, \zeta_3, \xi_3, \zeta_4, \xi_4,$ $\zeta_5, \xi_5, \zeta_6, \xi_6$ space. In general there are $2^{12} = 4096$ solutions to such a system of equations. Homotopy based polynomial continuation methods [138] may be used to find all the solutions to such a system of equations.

Of these solutions, only the ones with $\zeta_i, \xi_i$ such that the contact point is contained in the face, are the physically meaningful. Substituting those values back into Equations (I.67) and (I.68), we can find the rotation quaternion $q$, as described in Section H.3¶. This value of rotation may be substituted into Equations (I.69)-(I.78), to verify that it is valid.

Substituting the value of $q$ and $\zeta_i, \xi_i$ back into Equations (I.61)-(I.65) the gives the value of $\vec{t}$. For numerical stability, one can average the values obtained from each of these equations.

Note that the process of elimination can introduce "extra" solutions, especially since we have 12 unknowns and 48 equations. There may be other ways to solving this CDFS system of equations, which may result in a smaller number of solutions. Also, further reduction of the quadratic Equations (I.79)-(I.90) may be possible. It is also possible to directly solve Equations (I.67)-(I.78) as a the polynomial system of equations, using homotopy based polynomial continuation methods [138].

---

¶Note that Equations (I.67)-(I.78) have the same form as Equation (H.2).

# Bibliography

[1] AAAI. *The Theory and Applications of Minimal-Length Encoding*, AAAI Spring Symposium Series, Stanford University, March 1990. Working Notes.

[2] Mongi A. Abidi and Rafael C. Gonzalez, editors. *Data Fusion in Robotics and Machine Intelligence*. Academic Press, Inc., 1992.

[3] H. Akaike. A new look at the statistical model identification. *IEEE Transactions on Automatic Control*, AC-19:716–723, 1974.

[4] Michael S. Ali and Charles Engler, Jr. System description document for the Anthrobot-2: A dextrous robot hand. Technical Memorandum 104535, NASA Goddard Space Flight Center, Maryland, March 1991.

[5] Peter K. Allen. Integrating vision and touch for object recognition tasks. *The International Journal of Robotics Research*, 7(6):15–33, December 1988.

[6] Peter K. Allen and Kenneth S. Roberts. Haptic object recognition using a multi-fingered dextrous hand. In *Proc. 1989 IEEE International Conference on Robotics and Automation*, pages 342–347, Scottsdale, Arizona, 1989.

[7] Thomas Bäck, Frank Hoffmeister, and Hans-Paul Schwefel. A survey of evolution strategies. In *Proceedings of the Fourth International Conference on Genetic Algorithms*, pages 2–9, Los Altos, CA, 1991. Morgan Kauffman.

[8] Thomas Bäck and Hans-Paul Schwefel. An overview of evolutionary algorithms for parameter optimization. *Evolutionary Computation*, 1(1):1–23, 1993.

[9] Henry S. Baird. *Model-Based Image Matching Using Location*. MIT Press, Cambridge, Massachusetts, 1985.

[10] L. Banta and K.D. Rawson. Sensor fusion for mining robots. *IEEE Transactions on Industry Applications*, 30(5):1321–1325, 1994.

[11] A. J. Barbera, M. L. Fitzgerald, J. S. Albus, and L. S. Haynes. RCS: The NBS real-time control system. In *Robotics 8 Conference Proceedings*, pages 19.1–19.33, Detroit, 1984.

[12] A. Barron, J. Rissanen, and Bin Yu. The minimum description length principle in coding and modeling. *IEEE Transactions on Information Theory*, 44(6):2743–2760, October 1998.

[13] Andrew R. Barron and Thomas M. Cover. Minimum complexity density estimation. *IEEE Transactions on Information Theory*, 37(4):1034–1054, July 1991.

[14] O. A. Basir and H. C. Shen. Modeling and fusing uncertain multi-sensory data. *Journal of Robotic Systems*, 13(2):95–109, 1996.

[15] James O. Berger. *Statistical Decision Theory and Bayesian Analysis*. Springer Series in Statistics. Springer-Verlag, second edition, 1985.

[16] H. Bischof and A. Leonardis. Finding optimal neural networks for land use classification. *IEEE Transactions on Geoscience and Remote Sensing*, 36 (1):337–341, January 1998.

[17] R. E. Blahut. *Principles and Practice of Information Theory*. Addison-Wesley, second edition, 1990.

[18] I. Bloch. Information combination operators for data fusion: A comparative review with classification. *IEEE Transactions on Systems, Man and CyberneticsPart A*, 26(1):52–67, January 1996.

[19] Ruud M. Bolle and David B. Cooper. On optimally combining pieces of information, with application to estimating 3-D complex-object position from range data. *IEEE Transactions on Pattern Analysis and Machine Intelligence*, PAMI-8(5):619–638, September 1986.

[20] R. C. Bolles and R. A. Cain. Recognizing and locating partially visible objects: The local feature focus method. *The International Journal of Robotics Research*, 1(3):57–82, 1982.

[21] Grady Booch. *Object-Oriented Analysis and Design with Applications*. The Benjamin/Cummings Publishing Company, Inc., second edition, 1994.

[22] Francois Bourgeois and Jean-Claude Lassalle. An extension of the Munkres algorithm for the assignment problem to rectangular matrices. *Communications of the ACM*, 14(12):802–806, December 1971.

[23] Rafael Bracho, John F. Schlag, and Arthur C. Sanderson. POPEYE: A grey-level vision system for robotics applications. Technical Report CMU-RI-TR-83-6, Robotics Institute, Carnegie-Mellon University, Pittsburgh, Pennsylvania, May 1983.

[24] Thomas M. Breuel. Fast recognition using adaptive subdivisions of transformation space. In *Proc. IEEE Computer Society Conference on Computer Vision and Pattern Recognition*, pages 445–451, Champaign, Illinois, June 1992. IEEE Computer Society, Computer Society Press.

[25] R. A. Brooks. Symbolic reasoning around 3-D models and 2-D images. *Artificial Intelligence*, 17:285–348, 1981.

[26] Richard R. Brooks and Sundararaja S. Iyengar. *Multi-Sensor Fusion: Fundamentals and Applications with Software*. Prentice Hall, 1998.

[27] Zi-Xing Cai. *Intelligent Control: Principles, Techniques and Applications*, volume 7 of *Series in Intelligent Control and Intelligent Automation*. World

Scientific, 1997.

[28] J. Canning. A minimum description length model for recognizing objects with variable appearances (the VAPOR model). *IEEE Transactions on Pattern Analysis and Machine Intelligence*, 16(10):1032–1036, October 1994.

[29] Todd A. Cass. A robust parallel implementation of 2D model-based recognition. In *Proceedings of the Computer Society Conference on Computer Vision and Pattern Recognition*, pages 879–884, Ann Arbor, Michigan, June 1988. IEEE Computer Society, Computer Society Press.

[30] Gregory J. Chaitin. A theory of program size formally identical to information theory. *Journal of the ACM*, 22(3):329–340, 1975.

[31] Gregory J. Chaitin. *Algorithmic Information Theory*. Cambridge tracts in theoretical computer science 1. Cambridge University Press, New York, 1990.

[32] Gregory J. Chaitin. *Information Randomness and Incompleteness : Papers on Algorithmic Information Theory*. World Scientific, New Jersey, 2nd edition, 1990.

[33] I. Chakravarty. *The Use of Characteristic Views as a Basis for Recognition of Three-Dimensional Objects*. PhD thesis, Electrical, Computer and Systems Engineering Department, Rensselaer Polytechnic Institute, Troy, New York, December 1982.

[34] C. H. Chen and A. C. Kak. A robot vision system for recognizing 3-D objects in low-order polynomial time. *IEEE Transactions on Systems, Man, and Cybernetics*, 19(6):1535–1563, November/December 1989.

[35] Jian-Guo Chen and Nirwan Ansari. Adaptive fusion of correlated local decisions. *IEEE Transactions on Systems, Man and CyberneticsPart C: Applications and Reviews*, 28(2):276–281, May 1998.

[36] Laure J. Chipman, Timothy M. Orr, and Lewis N. Graham. Wavelets and image fusion. In *Wavelet Applications in Signal and Image Processing III*, volume 1, pages 208–219, San Diego, 1995. SPIE.

[37] Albert C.S. Chung and Helen C. Shen. Integrating dependent sensory data. In *Proc. 1998 IEEE International Conference on Robotics and Automation*, volume 4, pages 3546–3551, 1998.

[38] James J. Clark and Alan L. Yuille. *Data Fusion for Sensory Information Processing Systems*. The Kluwer International Series in Engineering and Computer Science, Robotics: Vision, Manipulation and Sensors. Kluwer Academic Publishers, Boston, 1990.

[39] Simon Cooper and Hugh Durrant-Whyte. A Kalman filter model for GPS navigation of land vehicles. In *Proc. 1994 International Conference on Intelligent Robots and Systems*, volume 1. IEEE/RSJ, 1994.

[40] Thomas M. Cover and Joy A. Thomas. *Elements of Information Theory*. Wiley Series in Telecommunications. John Wiley and Sons, Inc., 1991.

[41] John J. Craig. *Introduction to Robotics: Mechanics and Control*. Addison-Wesley, second edition, 1989.

[42] J. L. Crowley and Y. Demazeau. Principles and techniques for sensor data

fusion. *Signal Processing*, 32(1/2):5–27, May 1993.

[43] Robert Cunningham. Segmenting binary images. *Robotics Age*, pages 4–19, July/August 1981.

[44] A. Curran and K. J. Kyriakopoulos. Sensor-based self-localization for wheeled mobile robots. In *Proc. 1993 IEEE International Conference on Robotics and Automation*, volume 1, pages 8–13, Atlanta, Georgia, 1993.

[45] B.V. Dasarathy. Sensor fusion potential exploitation-innovative architectures and illustrative applications. *Proceedings of the IEEE: Special Issues on Data Fusion*, 85(1):24–38, January 1997.

[46] Mohamed Dekhil and Thomas C. Henderson. Instrumented sensor system architecture. *The International Journal of Robotics Research*, 17(4):402–417, 1998.

[47] P. M. Djuric. Asymptotic map criteria for model selection. *IEEE Transactions on Signal Processing*, 46(10):2726–2735, October 1998.

[48] R. O. Duda and P. E. Hart. *Pattern Classification and Scene Analysis*. John Wiley, New York, 1973.

[49] Dan E. Dudgeon and Russell M. Mersereau. *Multidimensional Digital Signal Processing*. Prentice-Hall, Englewood Cliffs, NJ, 1984.

[50] Hugh F. Durrant-Whyte. Consistent integration and propagation of disparate sensor observations. *The International Journal of Robotics Research*, 6(3):3–24, Fall 1987.

[51] Hugh F. Durrant-Whyte. Sensor models and multisensor integration. In Kak and Chen [105], pages 303–312.

[52] R. O. Eason and R. C. Gonzalez. Least-squares fusion of multisensory data. In Abidi and Gonzalez [2], chapter 9, pages 367–413.

[53] A. Elfes. Using occupancy grids for mobile robot navigation and perception. *IEEE Computer*, 22(6):46–58, June 1989.

[54] R. E. Ellis. Planning tactile recognition paths in two and three dimensions. *The International Journal of Robotics Research*, 11(2), April 1992.

[55] H. R. Everett. *Sensors for Mobile Robots: Theory and Applications*. Wellesly, Massachusetts, 1995.

[56] O. D. Faugeras and M. Hebert. The representation, recognition, and locating of 3-D objects. *The International Journal of Robotics Research*, 5(3):27–52, Fall 1986.

[57] M. A. T. Figueiredo and J. M. N. Leitao. Unsupervised image restoration and edge location using compound gauss-markov random fields and the mdl principle. *IEEE Transactions on Image Processing*, 6(8):1089–1102, August 1997.

[58] Martin A. Fischler and Robert C. Bolles. Random sample consensus: A paradigm for model fitting with applications to image analysis and automated cartography. *Communications of the ACM*, 24(6):381–395, June 1981.

[59] David B. Fogel. *System Identification Through Simulated Evolution: A Machine Learning Approach to Modeling*. Ginn Press, Needham Heights,

MA, 1991.

[60] Nigel Foster. *Attributed Image Matching Using a Minimum Representation Size Criterion.* PhD thesis, Carnegie Mellon University, Pittsburgh, Pennsylvania, 1987.

[61] Pascal Fua and Andrew J. Hanson. Objective functions for feature discrimination: Theory. In *Proceedings: Image Understanding Workshop*, pages 443–459. Morgan Kaufman, 1989.

[62] K. Fukunaga. *Statistical Pattern Recognition.* Academic Press, San Diego, 2nd edition, 1990.

[63] Erich Gamma, Richard Helm, Ralph Johnson, and John Vlissides. *Design Patterns.* Addison-Wesley, 1995.

[64] Stuart Geman and Donald Geman. Stochastic relaxation, Gibbs distribution, and Bayesian restoration of images. *IEEE Transactions on Pattern Analysis and Machine Intelligence*, PAMI-6(6):721–741, November 1984.

[65] Z. Gigus, J. Canny, and R. Seidel. Efficiently computing and representing aspect graphs of polyhedral objects. *IEEE Transactions on Pattern Analysis and Machine Intelligence*, 15(6):542–551, June 1991.

[66] David E. Goldberg. *Genetic Algorithms in Search, Optimization, and Machine Learning.* Addison-Wesley, 1989.

[67] M. Gondran and M. Minoux. *Graphs and algorithms.* John Wiley and Sons Inc., New York, 1984.

[68] I. R. Goodman, Ronald P. S. Mahler, and Hung T. Nguyen. *Mathematics of Data Fusion.* Kluwer Academic, 1997.

[69] Steven G. Goodridge, Michael G. Kay, and Ren C. Luo. Multilayered fuzzy behavior fusion for real-time reactive and control of systems with multiple sensors. *IEEE Transactions on Industrial Electronics*, 43(3):387–394, June 1996.

[70] W. Eric L. Grimson and Daniel P. Huttenlocher. On the sensitivity of the Hough transform for object recognition. *IEEE Transactions on Pattern Analysis and Machine Intelligence*, 12(3):255–274, March 1990.

[71] W. Eric L. Grimson and Tomas Lozano-Pérez. Model based recognition and localization from sparse range or tactile data. *The International Journal of Robotics Research*, 3(3):3–35, Fall 1984.

[72] X. E. Gros. *NDT Data Fusion.* John Wiley, 1997.

[73] P. Grossmann. Multisensor data fusion. *GEC Journal of Technology*, 15 (1):27–37, 1998.

[74] G. D. Hager. *Task-Directed Sensor Fusion and Planning: A Computational Approach.* Kluwer, Norwell, Massachusetts, 1990.

[75] David L. Hall and Gerald Kasmala. Visual programming environment for multisensor data fusion. In *Proceedings of SPIE*, volume 2764, pages 181–187. The International Society for Optical Engineering, 1996.

[76] David L. Hall and James Llinas. An introduction to multisensor fusion. *Proceedings of the IEEE: Special Issues on Data Fusion*, 85(1):6–23, January 1997.

[77]  Robert M. Haralick and Linda G. Shapiro. *Computer and Robot Vision*, volume 2. Addison-Wesley, 1993.

[78]  C.J. Harris, A. Bailey, and T.J. Dodd. Multi-sensor data fusion in defence and aerospace. *Aeronautical Journal*, 102(1015):229–244, May 1998.

[79]  Y. Hel-Or and M. Werman. Pose estimation by fusing noisy data of different dimensions. *IEEE Transactions on Pattern Analysis and Machine Intelligence*, 17(2):195–201, February 1995.

[80]  T. C. Henderson and E. Shilcrat. Logical sensor systems. *Journal of Robotic Systems*, 1(2):169–193, 1984.

[81]  H. Hexmoor. Architectural issues for integration of sensing and acting modalities. In *Proc. 1998 IEEE International Symposium on Intelligent Control*, pages 319–324, September 1998.

[82]  J. J. Hopfield and D. W. Tank. "Neural" computation of decisions in optimization problems. *Biological Cybernetics*, 52:141–152, 1985.

[83]  R. Hummel and S. Zucker. On the foundations of relaxation labeling processes. *IEEE Transactions on Pattern Analysis and Machine Intelligence*, PAMI-5:267–287, 1983.

[84]  Seth A. Hutchinson and Avinash C. Kak. Planning sensing strategies in a robot work cell with multi-sensor capabilities. *IEEE Transactions on Robotics and Automation*, 5(6):765–783, December 1989.

[85]  Daniel P. Huttenlocher and Shimon Ullman. Recognizing solid objects by alignment with an image. *International Journal of Computer Vision*, 5(2): 195–212, 1990.

[86]  R. A. Iltis and K. L. Anderson. A consistent estimation criterion for multi-sensor bearings-only tracking. *IEEE Transactions on Aerospace and Electronic Systems*, 32(1):108–120, January 1996.

[87]  Kim Intaek and G. Vachtsevanos. Overlapping object recognition: A paradigm for multiple sensor fusion. *IEEE Robotics & Automation Magazine*, 5(3):37–44, September 1998.

[88]  Tomohiko Ishikawa, Shigeyuki Sakane, Tomomasa Sato, and Hideo T-sukune. Estimation of contact position between a grasped object and the environment based on sensor fusion of vision and force. In *Proc. 1996 IEEE Conference on Multisensor Fusion*, pages 116–123. IEEE/SICE/RSJ, December 1996.

[89]  S. S. Iyengar, D. N. Jayasimha, and D. Nadig. A versatile architecture for the distributed sensor integration problem. *IEEE Transactions on Computers*, 43(2):175–185, February 1994.

[90]  S. S. Iyengar and L. Prasad. A general computational framework for distributed sensing and fault-tolerant sensor integration. *IEEE Transactions on Systems, Man and Cybernetics*, 25(4):643–650, April 1995.

[91]  William H. Jefferys and James O. Berger. Ockham's razor and bayesian analysis. *American Scientist*, 80(1):64–72, Jan-Feb 1992. Published by Sigma Xi.

[92]  Rajive Joshi. Determining the optimal number of clusters using a minimal

representation size criterion. Unpublished Project Report, December 1991.

[93] Rajive Joshi. CTOS++: Writing CTOS applications in C++. CAT Technical Memorandum 1, Center for Advanced Technology in Automation and Robotics, Rensselaer Polytechnic Institute, Troy, New York, January 1993.

[94] Rajive Joshi. CTOS++: An environment for distributed object-oriented programming. Technical Report CAT-TR, Center for Advanced Technology in Automation and Robotics, Rensselaer Polytechnic Institute, Troy, New York, May 1994.

[95] Rajive Joshi. *A Minimal Representation Framework for Multisensor Fusion and Model Selection*. PhD thesis, Rensselaer Polytechnic Institute, Troy, New York, December 1996. Electrical, Computer, and Systems Engineering Department.

[96] Rajive Joshi and Michael Ali. Catechism. Technical Report CAT-TR, Center for Advanced Technology in Automation and Robotics, Rensselaer Polytechnic Institute, Troy, New York, October 1994.

[97] Rajive Joshi and Arthur C. Sanderson. Shape matching from grasp using a minimal representation size criterion. In *Proc. 1993 IEEE International Conference on Robotics and Automation*, volume 1, pages 442–449, Atlanta, Georgia, May 1993.

[98] Rajive Joshi and Arthur C. Sanderson. Model-based multisensor data fusion: A minimal representation approach. In *Proc. 1994 IEEE International Conference on Robotics and Automation*, volume 1, pages 477–484, San Diego, May 1994.

[99] Rajive Joshi and Arthur C. Sanderson. Multisensor fusion and unknown statistics. In *Proc. 1995 IEEE International Conference on Robotics and Automation*, volume 3, pages 2670–2677, Nagoya, Japan, May 1995.

[100] Rajive Joshi and Arthur C. Sanderson. Multisensor fusion and model selection using a minimal representation size framework. In *Proc. 1996 IEEE Conference on Multisensor Fusion*, pages 25–32, Washington DC, December 1996. IEEE/RSJ/SICE.

[101] Rajive Joshi and Arthur C. Sanderson. Experimental studies in multisensor fusion. In *Proc. 1997 International Conference on Advanced Robots*, Monterey, CA, July 1997.

[102] Rajive Joshi and Arthur C. Sanderson. Minimal representation multisensor fusion using differential evolution. In *Proc. 1997 IEEE International Symposium on Computational Intelligence in Robotics and Automation*, Monterey, CA, July 1997. IEEE.

[103] Rajive Joshi and Arthur C. Sanderson. Multisensor fusion of touch and vision using minimal representation size. In *Proc. 1997 International Conference on Intelligent Robots and Systems*, volume 3. IEEE/RSJ, 1997.

[104] Rajive Joshi and Arthur C. Sanderson. Minimal representation multisensor fusion using differential evolution. *IEEE Transactions on Systems, Man and Cybernetics—Part A: Systems and Humans*, 29(1):63–76, January 1999.

[105] Avi Kak and Su-Shing Chen, editors. *Spatial Reasoning and Multi-Sensor*

*Fusion, Proceedings of the 1987 Workshop*, St. Charles, Illinois, October 1987. AAAI, Morgan Kaufmann Publishers, Los Altos, California.

[106] Moshe Kam, Xiaoxun Zhu, and Paul Kalata. Sensor fusion for mobile robot navigation. *Proceedings of the IEEE: Special Issues on Data Fusion*, 85(1): 108–119, January 1997.

[107] Michael Kass, Andrew Witkin, and Demetri Terzopoulos. Snakes: Active contour models. In *IEEE First International Conference on Computer Vision*, pages 259–268, London, June 1987.

[108] R.J. Kenefic. An algorithm to partition dft data into sections of constant variance. *IEEE Transactions on Aerospace and Electronic Systems*, 34(3): 789–795, July 1998.

[109] Aftab Ali Khan and M.A. Zohdy. Genetic algorithm for selection of noisy sensor data in multisensor data fusion. In *Proc. 1997American Control Conference*, volume 4. IEEE, 1997.

[110] Whoi-Yul Kim and Avinash C. Kak. 3-D object recognition using bipartite matching embedded in discrete relaxation. *IEEE Transactions on Pattern Analysis and Machine Intelligence*, 13(3):224–251, March 1991.

[111] G. Kinoshita, Y. Ikhsan, and H. Osumi. Sensor fusion with aspect information of visual and tactual sensing. In *Proc. 1998 International Conference on Intelligent Robots and Systems*, pages 1046–1052, Victoria, B.C., Canada, 1998.

[112] M. Kokar and K. Kim. Review of multisensor data fusion architectures and techniques. In *Proc. 1993 IEEE International Symposium on Intelligent Control*, Chicago, Illinois, August 1993. IEEE.

[113] A. N. Kolmogorov. Logical basis of information theory and probability theory. *IEEE Transactions on Information Theory*, IT-14(5):662–664, 1968.

[114] M. Korenaga and M. Hagiwara. Modified genetic programming based on elastic artificial selection and improved minimum description length. In *Proc. IEEE International Conference on Systems, Man, and Cybernetics*, volume 3, pages 2348–2353, October 1998.

[115] John R. Koza. *Genetic Programming: On the Programming of Computers by Means of Natural Selection*. MIT Press, 1992.

[116] Sukhan Lee. Sensor fusion and planning with perception-action network. *Journal of Intelligent and Robotic Systems: Theory & Applications*, 19(3): 271–298, July 1997.

[117] T. C. M. Lee. Segmenting images corrupted by correlated noise. *IEEE Transactions on Pattern Analysis and Machine Intelligence*, 20(5):481–492, May 1998.

[118] Paul E. Lehner, Kathryn Blackmond Laskey, and Didier Dubois. An introduction to issues in higher order uncertainty. *IEEE Transactions on Systems, Man and Cybernetics*, 26(3):289–293, 1996.

[119] John J. Leonard and Christopher M. Smith. Sensor data fusion in marine robotics. In *Proc. 19977th International Offshore and Polar Engineering Conference*, volume 2, pages 100–106, 1997.

[120] Harry R. Lewis and Christos H. Papadimitriou. *Elements of the Theory of Computation*. Prentice-Hall, Inc., Englewood Cliffs, NJ, 1981.

[121] Wei-Chung Lin, Fong-Yuan Liao, Chen-Kuo Tsao, and Theresa Lingulta. A hierarchical multiple-view approach to three-dimensional object recognition. *IEEE Transactions on Neural Networks*, 2(1):84–92, January 1991.

[122] Seppo Linnainmaa, David Harwood, and Larry S. Davis. Pose determination of a three-dimensional object using triangle pairs. *IEEE Transactions on Pattern Analysis and Machine Intelligence*, 10(5):634–646, September 1988.

[123] R. C. Luo and M. G. Kay. Multisensor integration and fusion in intelligent systems. *IEEE Transactions on Systems, Man and Cybernetics*, 19(5):901–931, September/October 1989.

[124] R. C. Luo and M. G. Kay. Data fusion and sensor integration: State-of-the-art 1990s. In Abidi and Gonzalez [2], chapter 2, pages 7–135.

[125] R. C. Luo and M. G. Kay. *Multisensor Integration and Fusion for Intelligent Machines and Systems*. Ablex Publishing Company, USA, 1995.

[126] R. C. Luo, M. Lin, and R. S. Scherp. Dynamic multi-sensor data fusion system for intelligent robots. *IEEE Journal of Robotics and Automation*, RA-4(4):386–396, 1988.

[127] R. C. Luo, A. Shr, and C-Y. Hu. Multiagent based multisensor resource management system. In *Proc. 1998 International Conference on Intelligent Robots and Systems*, pages 1034–1039, Victoria, B.C., Canada, 1998.

[128] Ren C. Luo. Sensor technologies and microsensor issues for mechatronics systems. *IEEE/ASME Transactions on Mechatronics*, 1(1):39–49, March 1996.

[129] Zhi-Quan Luo and J. N. Tsitsiklis. Data fusion with minimal communication. *IEEE Transactions on Information Theory*, 40(5):1551–1563, September 1994.

[130] Shoichi Maeyama, Akihisa Ohya, and Shin'ichi Yuta. Positioning by tree detection sensor and dead reckoning for outdoor navigation of a mobile robot. In *Proc. 1994 IEEE Conference on Multisensor Fusion*. IEEE, 1994.

[131] Shoichi Maeyama, Akihisa Ohya, and Shin'ichi Yuta. Non-stop outdoor navigation of a mobile robot. In *Proc. 1995 International Conference on Intelligent Robots and Systems*, volume 1, pages 130–135. IEEE/RSJ, 1995.

[132] James Manyika and Hugh Durrant-Whyte. *Data Fusion and Sensor Management: A Decentralized Information-Theoretic Approach*. Ellis Horwood, 1994.

[133] Keith Marzullo. Tolerating failures of continuous-valued sensors. *ACM Transactions on Computer Systems*, 8(4):284–304, November 1990.

[134] F. Matia and A. Jimenez. Multisensor fusion: An autonomous mobile robot. *Journal of Intelligent and Robotic Systems: Theory & Applications*, 22(2):129–141, June 1998.

[135] R. McKendall and M. Mintz. Data fusion techniques using robust statistics. In Abidi and Gonzalez [2], chapter 5, pages 211–244.

[136] Zbigniew Michalewicz. *Genetic Algorithms + Data Structures = Evolution Programs.* Springer, 1996.

[137] Micro Switch, Sensing and Control, Honeywell Inc., 11 West Spring Street, Freeport, Illinois 61032. *Basic Switches: Catalog 10*, May 1995.

[138] Alexander Morgan. *Solving Polynomial Systems Using Continuation for Engineering and Scientific Problems.* Prentice-Hall, Englewood Cliffs, New Jersey, 1987.

[139] J. L. Mundy and A. J. Heller. The evolution and testing of a model-based object recognition system. In *Third International Conference on Computer Vision: Proceedings*, pages 268–282, Osaka, Japan, December 1990. IEEE Computer Society.

[140] Robin R. Murphy. Dempster-shafer theory for sensor fusion in autonomous mobile robots. *IEEE Transactions on Robotics and Automation*, 14(2): 197–206, April 1998.

[141] Arthur G. O. Mutambara. *Decentralized Estimation and Control for Multisensor Systems.* CRC Press, Boca Raton, Florida, 1998.

[142] A. Najmi, R. A. Olshen, and R. M. Gray. A criterion for model selection using minimum description length. In B. Carpentieri, A. De Santis, U. Vaccaro, and J. A. Storer, editors, *Proceedings of Compression and Complexity of Sequences*, pages 204–214. IEEE, June 1998.

[143] Y. Nakamura. Geometric fusion: Minimizing uncertainty ellipsoid volumes. In Abidi and Gonzalez [2], chapter 11, pages 457–479.

[144] Akio Namiki and Masatoshi Ishikawa. Optimal grasping using visual and tactile feedback. In *Proc. 1996 IEEE Conference on Multisensor Fusion*, pages 589–596, IEEE/SICE/RSJ, December 1996.

[145] N. Nandhakumar and J. K. Aggarwal. Physics-based integration of multiple sensing modalities for scene interpretation. *Proceedings of the IEEE: Special Issues on Data Fusion*, 85(1):147–163, January 1997.

[146] Nasser M. Nasrabadi and Chang Y. Choo. Hopfield network for stereo vision correspondence. *IEEE Transactions on Neural Networks*, 3(1):5–13, January 1992.

[147] Nasser M. Nasrabadi and Wei Li. Object recognition by a Hopfield neural network. *IEEE Transactions on Systems, Man and Cybernetics*, 21(6): 1523–1535, November/December 1991.

[148] C.L. Nelson and D.S. Fitzgerald. Sensor fusion for intelligent alarm analysis. *IEEE Aerospace and Electronics Systems Magazine*, 12(9):18–24, September 1997.

[149] J. Russell Noseworthy. Inaccuracies in three dimensional vision systems. Master's thesis, Rensselaer Polytechnic Institute, Troy, New York, August 1991.

[150] J. Nygards and A. Wernersson. On covariances for fusing laser rangers and vision with sensors onboard a moving robot. In *Proc. 1998 International Conference on Intelligent Robots and Systems*, pages 1053–1059, Victoria, B.C., Canada, 1998.

[151] Athanasios Papoulis. *Probability, Random Variables, and Stochastic Processes.* McGraw-Hill, 3rd edition, 1991.

[152] L. Parida, D. Geiger, and R. Hummel. Junctions: Detection, classification, and reconstruction. *IEEE Transactions on Pattern Analysis and Machine Intelligence,* 20(7):687–698, July 1998.

[153] B. Parvin and G. Medioni. A layered network for the correspondence of 3-D objects. In *Proceedings of the 1991 IEEE International Conference on Robotics and Automation,* volume 2, pages 1808–1813, Sacramento, California, April 1991. IEEE Robotics and Automation Society.

[154] S.M.C. Peers. Blackboard system approach to electromagnetic sensor data interpretation. *Expert Systems,* 15(3):197–215, August 1998.

[155] John Porill. Optimal combination and constraints for geometrical sensor data. *The International Journal of Robotics Research,* 7(6):66–77, December 1988.

[156] Kenneth Price. Differential evolution: A fast and simple numerical optimizer. In *Proc. 1996North American Fuzzy Information Processing Society Conference,* 1996.

[157] Proceedings of the IEEE: Special issue on data fusion, January 1997.

[158] B. Ravichandran. *2D and 3D Model-Based Matching Using a Minimum Representation Criterion and a Hybrid Genetic Algorithm.* PhD thesis, Rensselaer Polytechnic Institute, Troy, New York, 1993.

[159] B. Ravichandran and A. C. Sanderson. Model based matching using simulated annealing and a minimum representation size criterion. In *Proceedings of the SPIE Symposium on Cooperative Intelligent Robotics in Space II,* volume 1612, pages 150–160, 1991.

[160] B. Ravichandran and A. C. Sanderson. Model based matching using a minimum representation size criterion and a hybrid genetic algorithm. In *Proceedings of the SPIE Symposium on Intelligent Robots and Visual Communication,* volume 1827, pages 76–87, 1992.

[161] Real-Time Innovations, Inc., Sunnyvale, CA. *ControlShell, The Component-Based Real-Time Programming System: User's Manual, Version 6.0,* 1999.

[162] Elaine Rich and Kevin Knight. *Artificial Intelligence.* McGraw-Hill, second edition, 1991.

[163] John M. Richardson and Kenneth A. Marsh. Fusion of multisensor data. *The International Journal of Robotics Research,* 7(6):78–96, December 1988.

[164] Josh Van Riper. The kinematics of an anthropomorphic robot hand. Master's thesis, Rensselaer Polytechnic Institute, Troy, New York, March 1992.

[165] J. Rissanen. Modeling by shortest data description. *Automatica,* 14:465–471, 1978.

[166] Jorma Rissanen. A universal prior for integers and estimation by minimum description length. *The Annals of Statistics,* 11(2):416–431, 1983.

[167] Jorma Rissanen. Universal coding, information, prediction, and estimation.

*IEEE Transactions on Information Theory*, IT-30(4):629–636, July 1984.

[168] Jorma Rissanen. Stochastic complexity and modeling. *The Annals of Statistics*, 14(3):1080–1100, 1986.

[169] Jorma Rissanen. *Stochastic Complexity in Statistical Inquiry.* World Scientific, New Jersey, 1989.

[170] Jorma Rissanen, Terry P. Speed, and Bin Yu. Density estimation by stochastic complexity. *IEEE Transactions on Information Theory*, 38(2): 315–323, March 1992.

[171] O. Rogalla, M. Ehrenmann, and R. Dillmann. A sensor fusion approach for pbd. In *Proc. 1998 International Conference on Intelligent Robots and Systems*, pages 1040–1045, Victoria, B.C., Canada, 1998.

[172] Thomas A. Runkler. Model based sensor fusion with fuzzy clustering. In *Proc. 1998International Conference on Fuzzy Systems*, volume 2, pages 1377–1382, 1998.

[173] Arthur C. Sanderson and Nigel J. Foster. Attributed image matching using a minimum representation size criterion. In *Proc. 1989 IEEE International Conference on Robotics and Automation*, volume 1, pages 360–365, Scottsdale, Arizona, 1989.

[174] Stanley A. Schneider, Vincent W. Chen, Gerardo Pardo-Castellote, and Howard H. Wang. ControlShell: A software architecture for complex electromechanical systems. *The International Journal of Robotics Research*, 17 (4):360–380, April 1998.

[175] G. Schwarz. Estimating the dimension of a model. *Annals of Statistics*, 6 (2):461–464, 1978.

[176] Hans-Paul Schwefel. *Evolution and Optimum Seeking.* John Wiley, 1995.

[177] Jakub Segen and Arthur C. Sanderson. Model inference and pattern discovery by minimal representation method. Technical Report CMU-RI-TR-82-2, The Robotics Institute, Carnegie-Mellon University, Pittsburgh, Pennsylvania 15213, July 1981.

[178] Shashank Shekhar, Oussama Khatib, and Makoto Shimojo. Object localization with multiple sensors. *The International Journal of Robotics Research*, 7(6):34–44, December 1988.

[179] David M. Siegel. Finding the pose of an object in a hand. In *Proc. 1991 IEEE International Conference on Robotics and Automation*, volume 1, pages 406–411, Sacramento, California, April 1991.

[180] Harjit Singh. Joint control of an anthropomorphic robot hand. Master's thesis, Rensselaer Polytechnic Institute, Troy, New York, October 1993.

[181] IEEE transactions on systems, man and cybernetics: Special issue on higher order uncertainty, May 1996.

[182] Randall C. Smith and Peter Cheeseman. On the representation and estimation of spatial uncertainty. *The International Journal of Robotics Research*, 5(4):56–68, Winter 1986.

[183] R. J. Solomonoff. A formal theory of inductive inference. *Information and Control*, 7:1–22, 1964.

[184] Harold W. Sorenson. *Parameter Estimation : Principles and Problems.* Marcel Dekker, New York, 1980.

[185] Rafael Sorkin. A quantitative Occam's razor. *International Journal of Theoretical Physics*, 22(12):1091–1104, 1983.

[186] M.E. Stieber, E. Petriu, and G. Vukovich. Instrumentation architecture and sensor fusion for systems control. *IEEE Transactions on Instrumentation and Measurement*, 47(1):108–113, February 1998.

[187] George Stockman and Sei-Wang Chen. Detecting the pose of rigid objects: a comparison of paradigms. In Hua-Kuang Liu and Paul S Schenker, editors, *Optical and Digital Pattern Recognition*, Proc. SPIE 754, pages 107–116. SPIE, January 1987.

[188] George Stockman and Juan Carlos Estava. 3D object pose from clustering with multiple views. *Pattern Recognition Letters*, 3:279–286, July 1985.

[189] Rainer Storn. Differential evolution design of an IIR-filter. In *Proc. 1996 IEEE International Conference on Evolutionary Computation*, pages 268–273. IEEE, May 1996.

[190] Rainer Storn and Kenneth Price. Minimizing the real functions of the ICEC'96 contest by differential evolution. In *Proc. 1996 IEEE International Conference on Evolutionary Computation*, pages 842–844. IEEE, May 1996.

[191] Bjarne Stroustrup. *The C++ Programming Language.* Addison-Wesley, second edition, 1992.

[192] Jayashree Subrahmonia, David B. Cooper, and Daniel Keren. Practical reliable bayesian recognition of 2D and 3D objects using implicit polynomials and algebraic invariants. LEMS Technical Report 107, Brown University, 1993.

[193] Gabor J. Tamasy. Smart sensor networks for robotic sensor skins. *Sensor Review*, 17(3):232–239, 1997.

[194] Demetri Terzopoulos, Andrew Witkin, and Michael Kass. Symmetry-seeking models for 3D object reconstruction. In *IEEE First International Conference on Computer Vision*, pages 269–276, London, June 1987.

[195] L. Thoraval, G. Carrault, J.M. Schleich, R. Summers, M. Van de Velde, and J. Diaz. Data fusion of electrophysiological and haemodynamic signals for ventricular rhythm tracking. *IEEE Engineering in Medicine and Biology Magazine*, 16(6):48–55, 1997.

[196] Pei-yih Ting and Ronald A. Iltis. Multitarget motion analysis in a DSN. *IEEE Transactions on Systems, Man and Cybernetics*, 21(5):1125–1139, September/October 1991.

[197] D. M. Titterington, A. F. M Smith, and U. E. Makov. *Statistical Analysis of Finite Mixture Distributions.* Wiley Series in Probability and Mathematical Statistics. Wiley, Chichester, New York, 1985.

[198] K. Umeda, T. Arai, and K. Ikushima. Fusion of range image and intensity image for 3D shape recognition. In *Proc. 1996 IEEE International Conference on Robotics and Automation*, volume 1, pages 690–685, 1996.

[199] Unimation, A Westinghouse Company, Unimation Incorporated, Shelter

Rock Lane, Danbury, CT 06810. *Unimate Puma Mark II Robot: 700 Series Equipment and Programming Manual*, April 1984.

[200] Unimation, A Westinghouse Company, Unimation Incorporated, Shelter Rock Lane, Danbury, CT 06810. *Unimate Industrial Robot Programming Manual: User's Guide to VAL II*, May 1985.

[201] Gert J. van Tonder and Johan J. Kruger. Shape encoding: A biologically inspired method of transforming boundary images into ensembles of shape-related features. *IEEE Transactions on Systems, Man and Cybernetics*, 27 (5):749–759, 1997.

[202] Pramod K. Varshney. *Distributed Detection and Data Fusion*. Springer, 1997.

[203] A. J. Vayda and A. C. Kak. A robot vision system for recognition of generic shaped objects. *Computer Vision, Graphics, and Image Processing-Image Understanding*, 54(1):1–46, July 1991.

[204] Virtual Technologies, 2175 Park Blvd, Palo Alto, CA 94306. *CyberGlove*, June 1993.

[205] Ellen L. Walker, Martin Herman, and Takeo Kanade. A framework for representing and reasoning about three-dimensional objects for vision. In Kak and Chen [105], pages 21–33.

[206] C. S. Wallace and D. M. Boulton. An information measure for classification. *Computer Journal*, 11(2):185, 1968.

[207] M. Wax, J. Sheinvald, and A. J. Weiss. Detection and localization in colored noise via generalized least squares. *IEEE Transactions on Signal Processing*, 44(7):1734–1743, July 1996.

[208] William M. Wells III. MAP model matching. In *Proc. IEEE Computer Society Conference on Computer Vision and Pattern Recognition*, pages 486–492. IEEE Computer Society, 1991.

[209] Peter Wide, Fredrik Winquist, Pontus Bergsten, and Emil M. Petriu. Human-based multi-sensor fusion method for artificial nose and tongue sensor data. In *Proc. 1998 IEEE Instrumentation and Measurement Technology Conference*, volume 1, pages 531–536, 1998.

[210] Man Leung Wong, Wai Lam, and Kwong Sak Leung. Using evolutionary programming and minimum description length principle for data mining of bayesian networks. *IEEE Transactions on Pattern Analysis and Machine Intelligence*, 21(2):174–178, February 1999.

[211] R.R. Yager. General approach to the fusion of imprecise information. *International Journal of Intelligent Systems*, 12(1):1–29, January 1997.

[212] Yasuharu Yamada, Akio Ishiguro, and Yoshiki Uchikawa. A method of 3D object reconstruction by fusing vision with touch using internal models with global and local deformations. In *Proc. 1993 IEEE International Conference on Robotics and Automation*, volume 2, pages 782–787, Atlanta, Georgia, May 1993.

[213] K. Yamanishi. A decision-theoretic extension of stochastic complexity and its applications to learning. *IEEE Transactions on Information Theory*, 44

(4):1424–1439, July 1998.

[214] Alan L. Yuille, Peter W. Hallinan, and David S. Cohen. Feature extraction from faces using deformable templates. *International Journal of Computer Vision*, 8(2):99–111, 1992.

[215] H. Zheng and S. D. Blostein. Motion-based object segmentation and estimation using the mdl principle. *IEEE Transactions on Image Processing*, 4(9):1223–1235, September 1995.

[216] Yifeng Zhou, H. Leung, and P.C. Yip. An exact maximum likelihood registration algorithm for data fusion. *IEEE Transactions on Signal Processing*, 45(6):1560–1573, June 1997.

[217] Yifeng Zhou and Henry Leung. Minimum entropy approach for multisensor data fusion. In *Proc. 1997IEEE Signal Processing Workshop on Higher-Order Statistics*, pages 336–339, 1997.

[218] Song Chun Zhu and A. Yuille. Region competition: Unifying snakes, region growing, and bayes/mdl for multiband image segmentation. *IEEE Transactions on Pattern Analysis and Machine Intelligence*, 18(9):884–900, September 1996.

[219] Ascension Technology Corporation, Burlington, Vermont. *The Bird$^{TM}$: Position and Orientation Measurement System*, November 1991.

# Index